Lecture Notes in Statistics

Edited by J. Berger, S. Fienberg, J. Gani, K. Krickeberg, and B. Singer

47

Albert J. Getson
Francis C. Hsuan

{2}-Inverses and Their Statistical Application

Springer-Verlag
New York Berlin Heidelberg London Paris Tokyo

Authors

Albert J. Getson
Merck, Sharp and Dohme Research Laboratories
West Point, PA 19486, USA

Francis C. Hsuan
Department of Statistics
School of Business Administration, Temple University
Philadelphia, PA 19122, USA

AMS Subject Classification (1980): 15-04, 15A09, 15A63, 62F99, 62H10, 65U05

ISBN-13: 978-0-387-96849-0 e-ISBN-13: 978-1-4612-3930-7

DOI: 10.1007/978-1-4612-3930-7

2847/3140-543210

To my daughters Joanne and Stephanie, ages 7 and 5,

who believe GAG – G has no meaning.

To Kathy, thanks for all your love through the years.

A.J.G.

PREFACE

Much of the traditional approach to linear model analysis is bound up in complex matrix expressions revolving about the usual generalized inverse. Motivated by this important role of the generalized inverse, the research summarized here began as an interest in understanding, in geometric terms, the four conditions defining the unique Moore-Penrose Inverse. Such an investigation, it was hoped, might lead to a better understanding, and possibly a simplification of, the usual matrix expressions.

Initially this research was begun by Francis Hsuan and Pat Langenberg, without knowledge of Kruskal's paper published in 1975. This oversight was perhaps fortunate, since if they had read his paper they may not have continued their effort. A summary of this early research appears in Hsuan, Langenberg and Getson (1985).

This monograph is a summary of the research on {2}-inverses continued by Al Getson, while a graduate student, in collaboration with Francis Hsuan of the Department of Statistics, School of Business Administration, at Temple University, Philadelphia.

The literature on generalized inverses and related topics is extensive and some of what is present here has appeared elsewhere. Generally, this literature is not presented from the point of view of {2}-inverses. We have tried to do justice to the relevant published works and appologize for those we have either overlooked or possibly misrepresented.

While it is our intention here to present a comprehensive study of {2}-inverses in statistics, we feel that this work is be no means exhaustive. Much work remains to be done, particularly in the area of multivariate analysis.

We wish to thank Dr. D. Raghavarao, Chairman of the Statistics Department at Temple University, for his encouragement of the publication of this work. We also thank the editorial staff at Springer-Verlag for their comments and suggestions on the preparation of the manuscript.

Finally the first author would like to thank his friends and colleagues in CBARDS at Merck, Sharp and Dohme Research Laboratories for the support he received in completing his research and degree.

A.J.G, F.C.H.
Philadelphia, PA
June 1988

TABLE OF CONTENTS

CHAPTER I
INTRODUCTION

A {2}-inverse for a given matrix A is any matrix G satisfying the second of the four conditions defining the unique Moore-Penrose Inverse of A:

$$(1)\ AGA = A \tag{1.1}$$

$$(2)\ GAG = G \tag{1.2}$$

$$(3)\ (AG)' = AG \tag{1.3}$$

$$(4)\ (GA)' = GA. \tag{1.4}$$

It is possible to construct matrices satisfying only a specified subset of the above conditions, for example (i),(j),...,(k). Such matrices, known as {i,j,...,k}-inverses, will be denoted $A^{+}_{i,j,\ldots,k}$. In this notation A^{+}_{1} is the usual g-inverse. Other classes of generalized inverses have been proposed in the literature and a number of texts have treated the subject in considerable depth. These include Pringle and Rayner (1971), Rao and Mitra (1971), and Ben-Israel and Greville (1974). In these works, the focus is generally on the {1}-inverse. In contrast, {2}-inverses, as their name implies, remain the stepchild of the {1}-inverse despite their importance in numerical analysis and electrical network theory [Ben-Israel and Greville (1974), pp. 27, 76].

The main function of the {1}-inverse is in solving a system of linear equations, especially when the system has deficiencies in rank, or is plainly inconsistent. It is our intention here to provide a comprehensive study of the {2}-inverse: its geometric characterization, algebraic properties, and uses in statistics. As we shall demonstrate, the {2}-inverse has several additional uses ranging from characterizing quadratic forms to computing algorithms in linear models.

When it comes to their applications in statistics, {2}-inverses are ubiquitous but not indispensible. In the statistical literature, {2}-inverses have had only limited exposure. As symmetric {1,2}-inverses, their role in least squares estimation was explored by Mazumdar et al. (1980), and by

Searle (1984). They have also been mentioned in connection with quadratic forms by several authors, including Carpenter (1950), Khatri (1963, 1977) and Mitra (1968). However, in these works the results were not viewed specifically in terms of {2}-inverses and the actual importance of {2}-inverses was clouded.

One reason for focusing on the {2}-inverses is that they provide an elegant mathematical language to express many ideas and results which otherwise involve cumbersome and laborious matrix expressions. This monograph contains numerous examples illustrating this simplicity. In this respect it is analogous to comparing different levels of computer programming languages. Assembly language is powerful but cumbersome. Even for a simple task such as counting from one to one hundred requires a long series of statements. On the other hand, a similar program in a higher level language such as BASIC, FORTRAN , or APL requires only a few statements. Currently, most textbooks in linear models contain complex matrix expressions. The language of {2}-inverses makes the expressions much simpler and, as a consequence, makes the underlying concepts much more transparent.

This work is organized into five chapters beginning with this Introduction. In each of the next four chapters a different aspect of {2}-inverses is explored. Consequently each chapter is somewhat, but not totally, independent of the others. The first section of each chapter introduces the problem considered there, reviews the relevant literature, and contains a detailed outline of the chapter. To aid the reader in understanding how these chapters relate, the following overview is offered.

A many-to-one mapping does not have an inverse in the traditional sense. In Chapter II, a functional approach is given for constructing generalized inverses of such a mapping, The Three Phase Inversion Procedure. As applied to a linear mapping over a real vector space, the procedure makes the {2}-inverse the focal point for a series of definitions of generalized inverses which is broader than the traditional. These include not only the usual g-inverses but also projection operators and constrained inverses. Such a {2}-inverse approach provides a conceptual framework unifying these various notions. This is due in part to the natural association {2}-inverses have with certain well defined vector spaces. The chapter continues with an investigation of several specific properties of {2}-inverses which suggest their usefulness in statistics. The first of these deals with the decomposition of a {2}-inverse into a sum of {2}-inverses of lesser rank and the joint decomposition of a pair of matrices into a weighted sum of {2}-inverses. Both of these results will be used in Chapter III to establish {2}-inverses as the natural coefficient matrix in a quadratic form assuring a χ^2 distribution. Next, as a generalization of projection operators, {2}-inverses are shown to have a key role in solving a system of linear equations. This role is further explored in Chapter IV in connection with least squares. Chapter II concludes with an algorithm for the easy computation of any {2}-inverse.

Given $\underline{x} \sim \mathcal{N}(\underline{\mu},\Sigma)$ with Σ nonsingular, then the well known necessary and sufficient condition for the χ^2 distribution of $\underline{x}'A\underline{x}$ can be restated to require A to be a {2}-inverse of Σ. In Chapter III, it is argued that the χ^2 distribution of $\underline{x}'A\underline{x}$ naturally forces A to be a {2}-inverse of Σ. With this as a basis, it is shown that for Σ singular, or otherwise, it is possible to represent, in a canonical way, all χ^2 distributed quadratic forms as $\underline{x}'A\underline{x}$, where A is a {2}-inverse of Σ. Quadratic forms are a special case of second degree polynomials, $\underline{x}'A\underline{x}+2\underline{b}'\underline{x} + c$. The distribution of second degree polynomials has been given by Khatri (1977); however, his approach for obtaining the parameters of the distribution is computationally difficult. By an application of the joint {2}-inverse decomposition of a pair of matrices, the notion of a canonical representation is expanded to include all second degree polynomials. In terms of this canonical representation, the parameters of the distribution may be easily expressed, and independent polynomials or polynomials following a χ^2 distribution can be readily identified.

In solving for least squares solutions (LSS's), a common approach is to assume a set of nonestimable constraints in addition to the normal equations so that a unique LSS may be found. The use of nonestimable constraints is usually viewed as a matter of convenience but not as a matter of necessity. On the other hand, the use of a g-inverse is viewed as necessary and sufficient to obtain a solution. In Chapter IV it is shown that, whether one realizes it or not, by choosing a g-inverse to obtain a solution, one is in fact imposing a set of constraints on the LSS. In particular, it is shown that every LSS may be expressed in the usual way as $\hat{\underline{\beta}} = GX'\underline{y}$ where X is the design matrix and G is a symmetric {1,2}-inverse of X'X. In turn each symmetric {1,2}-inverse is uniquely associated with a set of nonestimable constraints. Expanding upon this result, it is shown that for any set of constraints, regardless of their rank or estimability, the corresponding constrained LSS's may be found in an analogous way by an appropriate choice of a {2}-inverse of X'X. By appropriately identifying a set of constaints, the constrained LSS with the smallest norm may be easily identified. Since {2}-inverses may be easily calculated, the approach advocated in Chapter IV leads to a computational algorithm for obtaining LSS's. This approach does not depend on factorization of X or on solving for the eigenvalues of X'X. Furthermore, the approach is easily extended to weighted least squares.

For a designed experiment with an equal number of replications in each cell, there is little controversy concerning the sums of squares to be used in testing the various effects. However, when the data are imbalanced there is no concensus on what the appropriate sum of squares is for testing an effect. There are at least four alternative formulations of a linear model: the Unconstrained Model, the Constrained Model where the parameters are assumed to satisfy a set of known constraints, the Reparameterized Model where the constraints are used to reduce the model to one of full column rank, and the Cell Means Model in which a set of constraints is forced on the cell means to assure the equivalence of this model to some parameteric one.

Hypotheses may be expressed in terms of the parameters of one of these four models or in terms of the sample cell means alone. When there is imbalance, these five analytical approaches may each lead to a different numerator sum of squares for testing an effect. In Chapter V, the focus is on the development of an algorithm for identifying the hypotheses in each of the five approaches which results in algebraically identical numerator sum of squares. The algorithm, which is based upon {2}-inverses, is computationally simpler and covers a broader spectrum than other algorithms found in the literature. The algorithm is illustrated by its application to the SAS Type II and III Sums of Squares.

CHAPTER II
TIME FOR {2}-INVERSES

2.0 Introduction

As noted in the previous chapter, various classes of generalized inverses have been proposed in the literature. Geometric characterizations of generalized inverses were presented by Kruskal (1975) and, more recently, by Rao and Yanai (1985). The principal aim of this chapter is to unify and expand upon these diverse approaches in a consistent way.

The approach presented here begins with a geometric characterization of generalized inverses proposed by Hsuan, Langenberg and Getson (1985), the Three Phase Inversion Procedure. Their approach, which differs from the traditional, makes the {2}-inverse the natural starting point for a series of definitions of generalized inverses. The construction of various types of such generalized inverses are outlined. Included in this class are generalized inverses not defined entirely through the Moore-Penrose conditions. Some of these non-Penrose type inverses have statistical applications which will be explored in later chapters.

Of particular importance in statistics are symmetric {2}-inverses, a point of view defended in subsequent chapters. Symmetric {2}-inverses are a particular case of Bott-Duffin Inverses after a paper by Bott and Duffin who described their application in electrical network theory. {2}-inverses in general and Bott-Duffin Inverses in particular have several interesting properties and characterizations which will be useful in the following chapters and which are summarized in this chapter. The discussion of these begins with a few observations which suggest the role of Bott-Duffin Inverses in statistics. The discussion continues with an examination of the relationship between {2}-inverses and projection operators. This latter relationship leads to a decomposition theorem of symmetric matrices in terms of Bott-Duffin Inverses which is a generalization of the well known spectral decomposition. This chapter concludes with a discusssion of a procedure for efficient computation of any specified {2}-inverse.

This chapter is organized into eleven sections following this introduction. A brief description of the highlights of each section follows.

2.1 A functional definition of a generalized inverse is given in Definition 2.1 in terms of the Three Phase Inversion Procedure. Two types are identified: null and nonnull augmented generalized inverses.

2.2 The Three Phase Inversion Procedure is applied to linear mappings A: $\Re^n \mapsto \Re^m$. In Corollary 2.1.1, the constrained inverses of Rao are shown to be equivalent to the null and nonnull augmented generalized inverses of A.

2.3 The null augmented generalized inverses are identified to be exactly the class of {2}-inverses. In Theorem 2.2 and its Corollary 2.2.1, the correspondence between {2}-inverses and a pair of spaces $\mathcal{E} \subset \Re^n$ and $\mathcal{F} \subset \Re^m$ is established.

2.4 Theorem 2.3 describes the construction of any generalized inverse by augmenting a {2}-inverse in a nonnull way. As a particular case, the construction of the {1}-inverse, or usual g-inverse, is given in Corollary 2.3.1.

2.5 The construction of any Moore-Penrose Type generalized inverse by an appropriate choice of spaces $\mathcal{E}$ and $\mathcal{F}$ is detailed in Theorem 2.4.

2.6 The geometric relationships existing among the various subspaces associated with generalized inverses are summarized in Figure 2.1.

2.7 The usual projectors and their generalization by Rao are shown to be generalized inverses of the identity matrix in Lemma 2.5. As a converse, generalized inverses, which are not themselves projectors, have associated with themselves a pair of projectors. This association is outlined in Theorem 2.6.

2.8 The role of the usual g-inverse in solving a consistent set of linear equations is well known. As an extension, the role of generalized inverses in solving a broader system of equations is outlined in Theorem 2.7.

2.9 {2}-inverses can be decomposed into a sum of {2}-inverses of lesser rank. This decomposition, given in Lemma 2.8 and its corollaries, will be used repeatedly in subsequent chapters.

2.10 Theorem 2.9 details the Joint {2}-Inverse Decomposition of a pair of matrices, A and B, into a weighted sum of {2}-inverses. This decomposition may be viewed as a generalization of both the spectral and singular value decompositions. In Chapter III, the Joint {2}-Inverse Decomposition will lead to the characterization of the distribution of arbitrary quadratic forms.

2.11 {2}-inverses may be easily calculated. One approach, an application of the G2SWEEP operator, is discussed in this section.

2.1 The Three Phase Inversion Procedure

Classically the inverse of a mapping exists if and only if the mapping is bijective, i.e. one-to-one and onto. A many-to-one mapping $f: D \mapsto R$ does not have an inverse in the strict sense. Nevertheless, generalized inverses can be defined in terms of the Three Phase Inversion Procedure as follows:

1. The reduction phase, in which a subset D_0 of D is chosen such that f restricted to D_0 is bijective. Let the resulting mapping be denoted by $h: D_0 \mapsto R_0$.
2. The inversion phase, in which the unique inverse of h is determined, say $h^{-1}: R_0 \mapsto D_0$.
3. The augmentation phase, in which a mapping $g: D \mapsto R$ is defined so that $g = h^{-1}$ on R_0.

The resulting $g: R \mapsto D$ can be called a generalized inverse of f.

The nonuniqueness of a generalized inverse arises in two possible ways: the choice of D_0 in the reduction phase, and in the definition of g on the portion of the range space outside R_0 in the augmentation phase.

In practice, the choice of D_0 is not completely arbitrary, nor is the manner in which h^{-1} is augmented. For example, if D and R are vector spaces and f a linear mapping, it is natural to require in the reduction phase that D_0 be a subspace of D. Under this restriction, R_0 is a subspace of R and in the inversion phase h^{-1} is also a linear mapping of R_0 onto D_0. In the augmentation phase, h^{-1} may be extended to R in a number of ways. If R_1 is a complementary subspace of R_0 in R, then h^{-1} may be extended by either mapping R_1 trivially to the null vector or onto some nonnull space. The above discussion leads to the following definition.

Definition 2.1: For vector spaces D and R and a linear mapping $f: D \mapsto R$ then

1. a linear mapping $g: R \mapsto D$ is a Generalized Inverse of f, if there exists a subspace $D_0 \subset D$ such that $f: D_0 \mapsto f(D_0)$ is bijective, and
$$g \circ f(\underline{d}) = \underline{d} \text{ if and only if } \underline{d} \in D_0; \tag{2.1}$$
2. a generalized inverse is a Null Augmented Generalized Inverse if g maps some complementary subspace of $f(D_0)$ to the null vector;
3. a generalized inverse which is not null augmented is a Nonnull Augmented Generalized Inverse.

●

The above definition of generalized inverse is broader than the traditional one. It includes not only the usual g-inverse and other inverses defined through the Moore-Penrose conditions, but also the usual projectors and the projectors defined by Rao (1974).

Null and nonnull augmented generalized inverses have appeared in the guise of constrained inverses as defined by Rao and Mitra (1971). The relationship between constrained and generalized inverses is outlined in the next section.

2.2 Constrained Inverses

Although the Three Phase Inversion Procedure is quite general, attention will be focused on linear operators on real vector spaces. A real m×n matrix A of rank r defines two linear mappings,
$$A : \Re^n \mapsto \Re^m \tag{2.2}$$
and its transpose
$$A' : \Re^m \mapsto \Re^n. \tag{2.3}$$
For these mappings Rao and Mitra [(1971), p. 99] defined various Constrained Inverses, G, satisfying different combinations of the following constraints:

Type 1 Constraints

$$\text{c: } G \text{ maps vectors of } \Re^m \text{ into } \mathcal{S} \subset \Re^n \tag{2.4}$$
$$\text{r: } G' \text{ maps vectors of } \Re^n \text{ into } \mathcal{T} \subset \Re^m \tag{2.5}$$

Type 2 Constraints

$$\text{C: GA is an identity in } \mathcal{S} \tag{2.6}$$

$$\text{R: G'A' is an identity in } \mathcal{F}. \tag{2.7}$$

Table 2.1, on the next page, summarizes their results. It is not clear from an examination of Table 2.1, what relationships, if any, exist among the various constrained inverses. However, simple relationships do exist among these classes of inverses, which are easily seen through the Three Phase Inversion Procedure.

The existence of a generalized inverse G of A, as defined in the previous section, implies the existence of an s-dimension subspace $\mathcal{S} \subset \Re^n$ such that:

$$GA\underline{e} = \underline{e} \text{ if and only if } \underline{e} \in \mathcal{S}. \tag{2.8}$$

In what follows, it will be shown that (2.13) implies the existence of a unique s-dimensional subspace $\mathcal{F} \subset \Re^m$

$$G'A'\underline{f} = \underline{f} \text{ if and only if } \underline{f} \in \mathcal{F}. \tag{2.9}$$

Thus as a consequence, the Type 2 constraints (2.6) and (2.7) are equivalent.

Let $\mathcal{V}$ be the eigenspace of AG corresponding to the eigenvalue 1, then (2.13) implies

$$A(\mathcal{S}) \subset \mathcal{V}. \tag{2.10}$$

If $\underline{v} \in \mathcal{V}$, then

$$GAG\underline{v} = G\underline{v} \tag{2.11}$$

which in turn implies

$$G\underline{v} \in \mathcal{S} \tag{2.12}$$

and

$$G\underline{v} \neq \underline{0}. \tag{2.13}$$

Thus $\text{Dim}(\mathcal{V}) = s$, which implies $A(\mathcal{S}) = \mathcal{V}$. Since the eigenvalues of a matrix and its transpose are identical with the same multiplicity, the eigenspace of G'A' corresponding to the eigenvalue 1, $\mathcal{F} \subset \Re^m$, is nonempty with $\text{Dim}(\mathcal{F}) = s$. Thus for no larger space

$$G'A'\underline{f} = \underline{f} \text{ for all } \underline{f} \in \mathcal{F}. \tag{2.14}$$

Notice that G' is the generalized inverse of A' corresponding to the space $\mathcal{F}$.

A further relationship exists between $\mathcal{S}$ and $\mathcal{F}$. Since $\mathcal{F}$ and $A(\mathcal{S})$ are the left and right eigenspaces of AG corresponding to the eigenvalue 1, then

Table 2.1

Notation	Necessary and Sufficient Conditions For Existence	Algebraic Expression	
A_{cC}	$\mathrm{Rank}(AE) = \mathrm{Rank}(E)$	$(AE)^-$	(2.15)
A_{rR}	$\mathrm{Rank}(F'A) = \mathrm{Rank}(F)$	$(F'A)^-F'$	(2.16)
A_{cR}	$\mathrm{Rank}(F'AE) = \mathrm{Rank}(F)$	$E(F'AE)^-F' + E\left[I - F'AE(F'AE)^-F'AE\right]U$	(2.17)
A_{rC}	$\mathrm{Rank}(F'AE) = \mathrm{Rank}(E)$	$E(F'AE)^-F' + V\left[I - F'AE(F'AE)^-\right]F'$	(2.18)
A_{crCR}	$\mathrm{Rank}(F'AE) = \mathrm{Rank}(E) = \mathrm{Rank}(F)$	$E(F'AE)^-F'$	(2.19)

where $\mathcal{E} = \mathrm{Col}(E)$, $\mathcal{F} = \mathrm{Col}(F)$, U and V are arbitrary matrices, and $(AE)^-$, $(F'A)^-$ and $(F'AE)^-$ are arbitrary g-inverses.

$A(\mathcal{S})$ and $\mathcal{F}^{\perp}$, the Euclidean orthogonal complement of $\mathcal{F}$, are disjoint and together span $\Re^m$, i.e. $\Re^m = A(\mathcal{S}) \oplus \mathcal{F}^{\perp}$. Similarly, $\Re^n = A'(\mathcal{F}) \oplus \mathcal{S}^{\perp}$. This leads to the following lemma.

Lemma 2.1: Given two real matirces A, m×n, and G, n×m, for the existence of an s-dimensional subspace $\mathcal{S} \subset \Re^n$ such that

$$GA\underline{e} = \underline{e} \text{ if and only if } \underline{e} \in \mathcal{S} \tag{2.20}$$

it is necessary and sufficient that there exists an s-dimension subspace $\mathcal{F} \subset \Re^m$ such that

$$G'A'\underline{f} = \underline{f} \text{ if and only if } \underline{f} \in \mathcal{F}. \tag{2.21}$$

Under the conditions of the theorem

$$\mathcal{S} \cap \text{Null}(A) = \{\underline{0}\}, \tag{2.22}$$

$$\mathcal{F} \cap \text{Null}(A') = \{\underline{0}\}, \tag{2.23}$$

$$\Re^m = A(\mathcal{S}) \oplus \mathcal{F}^{\perp}, \tag{2.24}$$

and

$$\Re^n = A'(\mathcal{F}) \oplus \mathcal{S}^{\perp}. \tag{2.25}$$

●

As a consequence of Lemma 2.1, for subspaces $\mathcal{F}$ and $\mathcal{S}$ satisfying (2.22) through (2.25), the Type 2 constraints (2.6) and (2.7) are equivalent. This leads to the following corollary.

Corollary 2.1.1: As classes of matrices

$$A_{cC} = A_{rR} = A_{crCR} \subset A_{cR} = A_{rC} \tag{2.26}$$

Furthermore the matrices in A_{crCR} are null augmented generalized inverses and those in A_{cR} are nonnull augmented generalized inverses.

Proof: As discussed above, the Type 2 constraints are equivalent. Therefore (2.26) follows immediately. For any matrix G in A_{crCR}

$$\text{Rank}(G) = \text{Dim}(\mathcal{S}) = \text{Dim}(\mathcal{F}). \tag{2.27}$$

Consequently,

$$\text{Col}(G) = \mathcal{S} \tag{2.28}$$

and

$$\text{Row}(G) = \mathcal{F}. \tag{2.29}$$

However, (2.28) and (2.29) do not hold for all matrices in A_{cR}.

●

2.3 {2}- and {1,2}-Inverses: The Null Augmented Mappings

In terms of the Three Phase Inversion Procedure, conditions (2.22), (2.23), (2.24) and (2.25) have a natural interpretation. In the Reduction Phase, (2.22) and (2.23) ensure that

$$A : \mathcal{S} \mapsto A(\mathcal{S}), \tag{2.30}$$

and

$$A' : \mathcal{T} \mapsto A'(\mathcal{T}) \tag{2.31}$$

are both bijective. In the Inversion Phase, either (2.24) or (2.25), guarantees the existence of a matrix G, of rank s, satisfying (2.20) and (2.21), i.e. a null augmented generalized inverse. The following theorem is a variation of that appearing in Rao and Mitra [(1971), p. 101].

Theorem 2.2: For an $m \times n$ matrix A of rank r, and s-dimensional subspaces $\mathcal{S} \subset \Re^n$ and $\mathcal{T} \subset \Re^m$ satisfying (2.22), (2.23), (2.24) and (2.25), then the $n \times m$ matrix G of rank s satisfying

$$GA\underline{e} = \underline{e} \quad \text{if and only if } \underline{e} \in \mathcal{S} \tag{2.32}$$

and

$$G'A'\underline{f} = \underline{f} \quad \text{if and only if } \underline{f} \in \mathcal{T}, \tag{2.33}$$

may be uniquely expressed as

$$G = E(F'AE)^{-1}F', \tag{2.34}$$

where E and F are any full column rank matrices such that $\mathcal{S} = \text{Col}(E)$ and $\mathcal{T} = \text{Col}(F)$.

Proof:

($\Leftarrow$) Any G of the form (2.34) trivially satisfies (2.32) and (2.33).

($\Rightarrow$) Since $\text{Col}(G) = \mathcal{S}$ and $\text{Col}(G') = \mathcal{T}$, from the singular value decomposition of G

$$G = EKF' \tag{2.35}$$

where K is nonsingular. Thus

$$(2.32) \Rightarrow E = EKF'AE \Rightarrow I = KF'AE. \tag{2.36}$$

Similarly,

$$(2.33) \Rightarrow I = F'AEK. \tag{2.37}$$

Consequently,

$$K = (F'AE)^{-1} \tag{2.38}$$

and (2.34) follows.

●

A corollary of the above theorem, which appears in Ben-Israel and Greville [(1974), p. 60], is that all matrices satisfying the second of the

Moore-Penrose conditions are exactly those which can be expressed as (2.34)

Corollary 2.2.1: Given A, then

$$GAG = G \tag{2.39}$$

if and only if

$$G = E(F'AE)^{-1}F' \tag{2.40}$$

where $\mathcal{S} = Col(G)$, $\mathcal{F} = Col(G')$ with E and F are any full column rank matrices such that $\mathcal{S} = Col(E)$ and $\mathcal{F} = Col(F)$.

Proof: Trivial, since every {2}-inverse of A satisfies (2.32) and (2.33).

•

As will be shown in the next section, as a consequence of the above theorem and corollary, it is possible to construct matrices satisfying any specified set of the conditions (1.1), (1.2), (1.3) and (1.4). Such matrices are generalized inverses in the sense of the Definition 2.1. This leads to the following definition:

Definition 2.2: For a matrix A, a matrix G satisfying the Moore-Penrose conditions (i),(j),...,(k) is called an {i,j,...,k}-inverse of A and is denoted

$$A^{+}_{i,j,\cdots,k}. \tag{2.41}$$

It is also known as a Moore-Penrose Type Generalized Inverse.

•

As pointed out by Ben-Israel and Greville (1974), the above corollary establishes the correspondence between {2}-inverses of A, and spaces $\mathcal{S}$ and $\mathcal{F}$ satisfying (2.22), (2.23), (2.24) and (2.25).

Definition 2.3: Given an $m \times n$ matrix A

1. the unique {2}-inverse of A associated with $\mathcal{S} = Col(G)$ and $\mathcal{F} = Col(G')$ is denoted

$$A^{+}_{\mathcal{S},\mathcal{F}}; \tag{2.42}$$

2. if A is square, the Bott-Duffin inverse of A is $A^{+}_{\mathcal{S},\mathcal{S}}$ which is denoted

$$A^{+}_{\mathcal{S}} \tag{2.43}$$

•

The following corollary outlines a simple construction of {1,2}-inverses.

Corollary 2.2.2: Given an $m \times n$ matrix A of rank r and the {2}-inverse $G = A^{+}_{\mathcal{S},\mathcal{F}}$, then

$$AGA = A \tag{2.44}$$

if and only if

$$\mathrm{Rank}(G) = \mathrm{Rank}(A). \tag{2.45}$$

Proof:

($\Leftarrow$) If $\mathrm{Dim}(\mathcal{S}) = \mathrm{Dim}(\mathcal{T}) = \mathrm{Rank}(A)$ then in addition to (2.32) and (2.33)

$$AG\underline{x} = \underline{x} \quad \text{if and only if} \quad \underline{x} \in A(\mathcal{S}) \tag{2.46}$$

and $$A'G'\underline{y} = \underline{y} \quad \text{if and only if} \quad \underline{y} \in A'(\mathcal{T}). \tag{2.47}$$

Thus A is a {1,2}-inverse of G.

($\Rightarrow$) Trivial since Rank(A) = Rank(G).

●

{1,2}-inverses are also known as Reflexive Generalized Inverses.

2.4 {1.5}-Inverses: The Nonnull Augmented Mappings

In the previous section, null augmented generalized inverses were shown to be the class of {2}-inverses which includes the {1,2}-inverses. In this section, nonnull augmented generalized inverses are characterized.

For a nonnull augmented generalized inverse satisfying (2.20) and (2.21) the arbitrariness in the augmentation phase is provided for by an $n \times m$ matrix Z satisfying

$$ZAE = 0 \tag{2.48}$$

and $$Z'A'F = 0 \tag{2.49}$$

where E and F are full column rank matrices with $\mathrm{Col}(E) = \mathcal{S}$ and $\mathrm{Col}(F) = \mathcal{T}$. If $\Theta = [\underline{\theta}_1 \mid \cdots \mid \underline{\theta}_{m-s}]$ and $\Phi = [\underline{\phi}_1 \mid \cdots \mid \underline{\phi}_{n-s}]$ are full column rank matrices such that $\mathrm{Col}(\Theta) = [A(\mathcal{S})]^{\perp}$ and $\mathrm{Col}(\Phi) = [A'(\mathcal{T})]^{\perp}$, then Z may be expressed as

$$A = \sum_{i=1}^{n-s} \sum_{j=1}^{m-s} \lambda_{ij}\, \underline{\phi}_i\, \underline{\theta}_j' = \Phi\Lambda\Theta' \tag{2.50}$$

where $\Lambda = [\lambda_{ij}]$ is an arbitrary $(n\text{-}s) \times (m\text{-}s)$ matrix. Consequently, the matrices Z satisfying (2.48) and (2.49) form an $(n\text{-}s)(m\text{-}s)$ dimensional vector space. This implies that for given $\mathcal{S}$ and $\mathcal{T}$ there are $(n\text{-}s)(m\text{-}s)$ linearly independent choices for Z. The following theorem summarizes this characterization of generalized inverses.

Theorem 2.3: Given A, an m×n matrix, then an n×m matrix G is a generalized inverse of A such that for no larger s-dimensional spaces $\mathcal{S}$ and $\mathcal{F}$,

$$GA\underline{e} = \underline{e} \text{ for all } \underline{e} \in \mathcal{S} \tag{2.51}$$

and

$$G'A'\underline{f} = \underline{f} \text{ for all } \underline{f} \in \mathcal{F} \tag{2.52}$$

if and only if

$$G = A^{+}_{\mathcal{S},\mathcal{F}} + Z, \tag{2.53}$$

where Z is any linear mapping such that

$$Z\underline{g} = 0 \text{ for all } \underline{g} \in A(\mathcal{S}) \tag{2.54}$$

and

$$Z'\underline{h} = 0 \text{ for all } \underline{h} \in A'(\mathcal{F}). \tag{2.55}$$

In which case Rank(G) = Dim($\mathcal{S}$)+Rank(Z). In addition, $\mathcal{S}$ and $\mathcal{F}$ satisfy (2.22) to (2.25).

•

Given Dim($\mathcal{S}$) = Dim($\mathcal{F}$) = Rank(A), {1,2}-inverses result if, as seen in the previous section, $Z \equiv 0$. However, when Z is nonzero the resulting G is a {1}-inverse.

Corollary 2.3.1: For A, an m×n matrix of rank r, let $\mathcal{S} \subset \mathfrak{R}^n$ and $\mathcal{F} \subset \mathfrak{R}^m$ be r-dimensional subspaces satisfying (2.22), (2.23), (2.24) and (2.25). Consider the generalized inverses of the form

$$G = A^{+}_{\mathcal{S},\mathcal{F}} + Z \tag{2.56}$$

with Z satisfying ZA = 0 and AZ = 0, then

$$AGA = AA^{+}_{\mathcal{S},\mathcal{F}}A = A. \tag{2.57}$$

Conversely, every {1}-inverse of A may be uniquely expressed as in (2.56) with

$$A^{+}_{\mathcal{S},\mathcal{F}} = GAG \tag{2.58}$$

and

$$Z = G - GAG. \tag{2.59}$$

Proof: Trivial.

•

The characterization of {1}-inverses contained in the above corollary is not new. Rao and Yanai (1985) proved an identical result in characterizing *L*,*M*,N-inverses, their name for {1}-inverses. The result also appears as an exercise in Ben-Israel and Greville [(1974), p. 80].

If $\mathcal{S}$ and $\mathcal{F}$ are such that Dim($\mathcal{S}$) = Dim($\mathcal{F}$) < Rank(A) and $Z \not\equiv 0$, then G is not a Moore-Penrose type generalized inverse. They are relatives of the

{1}-inverses of A in the sense that they are nonnull augmented {2}-inverses instead of nonnull augmented {1,2}-inverses. Consequently, we may think of these inverses as falling between {1}- and {2}-inverses. In the spirit of {i,j,...,k}-inverses, we will call such inverses {1.5}-inverses. Particular cases of {1.5}-inverses have appeared in the literature. From Corollary 2.2.1, {1.5}-inverses are equivalent to the class A_{cR} of constrained inverses defined by Rao and Mitra (1971). The projectors defined by Rao (1974) are also members of this class. These points will be discussed in Section 2.7.

2.5 Construction of Moore-Penrose Type Generalized Inverses

It turns out that starting from {2}- or {1,2}-inverses, any Moore-Penrose type generalized inverse can be constructed by choosing the appropriate $\mathcal{S}$ and $\mathcal{F}$. The choice of $\mathcal{S}$ and $\mathcal{F}$ in the construction of other Moore-Penrose type generalized inverses is given in the following theorem.

Theorem 2.4:

Let G be a {2}-inverse of A with respect to $\mathcal{S}$ and $\mathcal{F}$, then

$$1.\ \mathcal{F} = A(\mathcal{S}) \text{ if and only if } (AG)' = AG \tag{2.60}$$

and

$$2.\ \mathcal{S} = A'(\mathcal{F}) \text{ if and only if } (GA)' = GA. \tag{2.61}$$

Let G be a {1}-inverse of A with respect to $\mathcal{S}$ and $\mathcal{F}$, written as

$$G = A^{+}_{\mathcal{S},\mathcal{F}} + Z,\ AZ = 0 \text{ and } ZA = 0 \tag{2.62}$$

then

$$3.\ \mathcal{F} = A(\mathcal{S}) \text{ if and only if } (AG)' = AG \tag{2.63}$$

and

$$4.\ \mathcal{S} = A'(\mathcal{F}) \text{ if and only if } (GA)' = GA. \tag{2.64}$$

Proof: Let E and F be full column rank matrices with Col(E) = $\mathcal{S}$ and Col(F) = $\mathcal{F}$, then

$$1.\ (\Rightarrow)\ AG = AE(F'AE)^{-1}F' = F(E'A'F)^{-1}E'A = G'A' = (AG)' \tag{2.65}$$

$$(\Leftarrow)\qquad (AG)' = AG \Rightarrow \begin{bmatrix} AGF = F \\ AGAE = AE \end{bmatrix} \Rightarrow A(\mathcal{S}) = \mathcal{F} \tag{2.66}$$

$$2.\ (\Rightarrow)\ GA = E(F'AE)^{-1}F'A = A'F(E'A'F)^{-1}E' = A'G' = (GA)' \tag{2.67}$$

$$(\Leftarrow)\qquad (GA)' = GA \Rightarrow \begin{bmatrix} GAE = E \\ GAA'F = A'F \end{bmatrix} \Rightarrow A'(\mathcal{F}) = \mathcal{S}. \tag{2.68}$$

3 and 4 follow trivially from 1 and 2 since AZ = 0 and ZA = 0.

●

Theorems 2.2, 2.3 and 2.4 make possible the algebraic construction of any Moore-Penrose type generalized inverse. As an example, the unique Moore-Penrose Inverse of A, denoted A^+, is given by

$$A^+ = E(F'AE)^{-1}F' \tag{2.69}$$

with E and F full column rank matrices such that Col(E) = Col(A′) and Col(F) = Col(A). Ben-Israel and Greville [(1974), p. 287] refer to (2.69) as Zlobec's formula.

The computation of {2}-inverses will be further discussed in Section 2.11.

2.6 A Geometric Representation of {2}-Inverses

Given a generalized inverse $G = A^+_{\mathcal{S},\mathcal{F}}+Z$ of the linear transformation $A: \mathfrak{R}^n \mapsto \mathfrak{R}^m$ consider the following subspaces:

$$\begin{aligned}
\mathcal{S} &= \left\{ \underline{x} \in \mathfrak{R}^n \mid GA\underline{x} = \underline{x} \right\}, & \mathcal{T} &= \left\{ \underline{y} \in \mathfrak{R}^m \mid AG\underline{y} = \underline{y} \right\},\\
\mathcal{S}_0 &= \left\{ \underline{x} \in \mathfrak{R}^n \mid GA\underline{x} = \underline{0} \right\}, & \mathcal{T}_0 &= \left\{ \underline{y} \in \mathfrak{R}^m \mid AG\underline{y} = \underline{0} \right\},\\
\mathcal{U} &= \left\{ \underline{x} \in \mathfrak{R}^n \mid \underline{x}'GA = \underline{x}' \right\}, & \mathcal{F} &= \left\{ \underline{y} \in \mathfrak{R}^m \mid \underline{y}'AG = \underline{y}' \right\},\\
\mathcal{U}_0 &= \left\{ \underline{x} \in \mathfrak{R}^n \mid \underline{x}'GA = \underline{0}' \right\}, & \mathcal{F}_0 &= \left\{ \underline{y} \in \mathfrak{R}^m \mid \underline{y}'AG = \underline{0}' \right\}.
\end{aligned} \tag{2.70}$$

From the discussion in Section 2.2 it is easily verified that the following relationships exist among these subspaces:

$$\mathrm{Dim}(\mathcal{S}) = \mathrm{Dim}(\mathcal{T}) = \mathrm{Dim}(\mathcal{U}) = \mathrm{Dim}(\mathcal{F}), \tag{2.71}$$

$$\mathrm{Dim}(\mathcal{S}_0) = \mathrm{Dim}(\mathcal{U}_0), \tag{2.72}$$

$$\mathrm{Dim}(\mathcal{T}_0) = \mathrm{Dim}(\mathcal{F}_0), \tag{2.73}$$

$$\mathcal{S} \perp \mathcal{U}_0, \tag{2.74}$$

$$\mathcal{U} \perp \mathcal{S}_0, \tag{2.75}$$

$$\mathcal{T} \perp \mathcal{F}_0, \tag{2.76}$$

$$\mathcal{F} \perp \mathcal{T}_0, \tag{2.77}$$

$$\mathrm{Null}(A) \subset \mathcal{S}_0, \tag{2.78}$$

$$\mathrm{Null}(A') \subset \mathcal{F}_0, \tag{2.79}$$

$$\mathcal{T} \subset \mathrm{Col}(A), \tag{2.80}$$

and

$$\mathcal{U} \subset \mathrm{Col}(A'). \tag{2.81}$$

In addition, the mappings

$$A : \mathcal{S} \mapsto \mathcal{T}, \tag{2.82}$$

$$G : \mathcal{T} \mapsto \mathcal{S}, \tag{2.83}$$

$$A' : \mathcal{F} \mapsto \mathcal{U}, \tag{2.84}$$

and

$$G' : \mathcal{U} \mapsto \mathcal{F} \tag{2.85}$$

are all bijective. In contrast, the mappings

$$A : \mathcal{S}_0 \mapsto \mathcal{V}_0, \tag{2.86}$$

$$G : \mathcal{V}_0 \mapsto \mathcal{S}_0, \tag{2.87}$$

$$A' : \mathcal{F}_0 \mapsto \mathcal{U}_0, \tag{2.88}$$

and

$$G' : \mathcal{U}_0 \mapsto \mathcal{F}_0, \tag{2.89}$$

although well defined, will not in general be bijections.

The relationships among these spaces may be conveniently summarized in the geometric representation detailed in Figure 2.1.

FIGURE 2.1

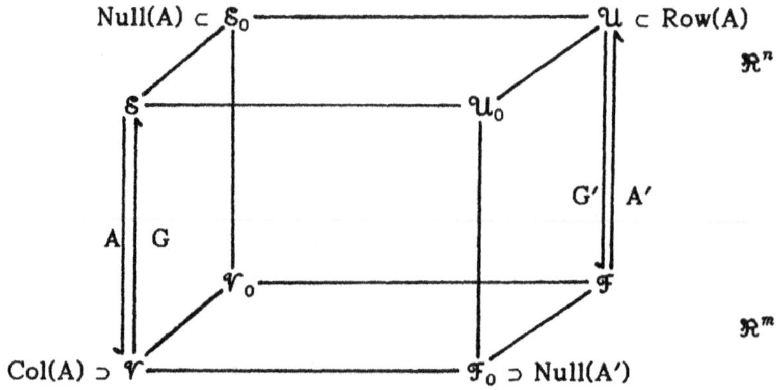

As also discussed in Section 2.2, G will be a {2}-inverse of A if and only if any one of the following equivalent conditions hold:

$$\mathcal{S} \oplus \mathcal{S}_0 = \Re^n, \tag{2.90}$$

$$\mathcal{U} \oplus \mathcal{U}_0 = \Re^n, \tag{2.91}$$

$$\mathcal{F} \oplus \mathcal{F}_0 = \Re^m, \tag{2.92}$$

$$\mathcal{V} \oplus \mathcal{V}_0 = \Re^m, \tag{2.93}$$

$$\mathcal{S}^{\perp} = \mathcal{U}, \tag{2.94}$$

$$\mathcal{U}^{\perp} = \mathcal{S}, \tag{2.95}$$

$$\mathcal{V}^{\perp} = \mathcal{F}, \tag{2.96}$$

or

$$\mathcal{F}^{\perp} = \mathcal{V}. \tag{2.97}$$

G will be a {1}-inverse of A if and only if

$$\text{Dim}(\mathcal{F}) = \text{Dim}(\mathcal{S}) = \text{Rank}(A). \tag{2.98}$$

In this case the following hold:

$$\mathcal{S}_0 = \text{Null}(A), \tag{2.99}$$

$$\mathcal{V} = \text{Col}(A), \tag{2.100}$$

$$\mathcal{F}_0 = \text{Null}(A'), \tag{2.101}$$

and

$$\mathcal{U} = \text{Col}(A'). \tag{2.102}$$

Equations (2.99) and (2.100) have appeared previous in the literature in a paper by Kruskal (1975) where he defined condition (1.1) as the "*sine qua non* for the notion of a generalized inverse." However, (1.1) is too restrictive in that if forces (2.98) to hold, which leads to a smaller class of generalized inverses than that defined through the more general Three Phase Inversion Procedure.

Finally, for G either a {1}- or a {2}-inverse of A, the results of Theorem 2.4 may be re-expressed in terms of the notation in Figure 2.1, as follows:

$$(AG)' = AG \text{ if and only if } \mathcal{F} = \mathcal{V}, \tag{2.103}$$

in which case

$$\mathcal{F}_0 = \mathcal{V}_0. \tag{2.104}$$

Similarly,

$$(GA)' = GA \text{ if and only if } \mathcal{S} = \mathcal{U}, \tag{2.105}$$

in which case

$$\mathcal{S}_0 = \mathcal{U}_0. \tag{2.106}$$

2.7 {1.5}-Inverses and Projections

For complementary subspaces Γ and Ψ of $\Re^n$, a linear mapping defined by

$$\wp\underline{\psi} = \underline{\psi} \quad \text{if and only if} \quad \underline{\psi} \in \Psi \tag{2.107}$$

and

$$\wp\underline{\gamma} = \underline{0} \quad \text{if and only if} \quad \underline{\gamma} \in \Gamma \tag{2.108}$$

is the traditional definition of a projection onto Ψ along Γ. From (2.107) it is clear that, in the sense of Definition 2.1, any such linear operator is a generalized inverse of the identity matrix. In fact, since $\wp$ is a projection if and only if it is idempotent, $\wp$ is a {2}-inverse of the identity matrix

$$\wp = I^{+}_{\Psi,\Gamma^{\perp}}. \tag{2.109}$$

Rao (1974) extended this notion of a projection operator to include any mapping $\wp$ satisfying (2.107) and (2.108), where Ψ and Γ are disjoint but not necessarily complementary subspaces. To give credit to Rao, where it is due, the following definition is made.

Definition 2.4:

For disjoint subspaces Γ and Ψ of $\Re^n$:

1. the Projector to Ψ along Γ is any linear mapping $\wp$ satisfying (2.107) and (2.108);
2. if $\Psi \oplus \Gamma = \Re^n$, then $\wp$ is the Full Rank Projector to Ψ along Γ which is denoted
$$\wp_{\Psi|\Gamma}; \tag{2.110}$$
3. if $\Psi \oplus \Gamma \neq \Re^n$, then $\wp$ is the Rao Projector to Ψ along Γ which is denoted
$$\wp^*_{\Psi|\Gamma}. \tag{2.111}$$

●

When $\Psi \oplus \Gamma$ is not the entire space, there are numerous Rao projectors satisfying (2.107) and (2.108) and $\wp^*_{\Psi|\Gamma}$ is not unique. Since Rao projectors are also generalized inverses of the identity matrix, Theorem 2.3 leads to the following lemma.

Lemma 2.5: For disjoint subspaces Γ and Ψ of $\Re^n$:

1. if $\Gamma \oplus \Psi = \Re^n$, then
$$\wp_{\Psi|\Gamma} = I^+_{\Psi,\Gamma^\perp}; \tag{2.112}$$
2. if $\Gamma \oplus \Psi \neq \Re^n$, then
$$\wp^*_{\Psi|\Gamma} = I^+_{\Psi,\Gamma_*^\perp} + Z, \tag{2.113}$$

   where $\Gamma \subset \Gamma_*$ with $\Re^n = \Psi \oplus \Gamma_*$ and Z is some linear mapping such that
$$Z\underline{g} = \underline{0} \text{ if and only if } \underline{g} \in \Psi \oplus \Gamma \tag{2.114}$$
   and
$$Z'\underline{h} = \underline{0} \text{ if and only if } \underline{h} \in \Gamma_*^\perp. \tag{2.115}$$

Proof:

1. Follows from (2.109).
2. From (2.111), Γ is the right eigenspace corresponding to 0, Ψ is the right eigenspace corresponding to 1 and let $\Gamma_*^\perp$ be the left eigenspace

corresponding to 1. Then

$$\Gamma_*^{\perp} \perp \Gamma, \tag{2.116}$$

and

$$\Gamma \subset \Gamma_*. \tag{2.117}$$

The result now follows from Theorem 2.3.

●

As with generalized inverses, when $\Gamma \oplus \Psi \neq \mathfrak{R}^n$ the nonuniqueness of $\wp^*_{\Psi|\Gamma}$ arises in two possible ways: in choice of Γ_* and in in the choice of the mapping Z satisfying (2.114) and (2.115).

A converse of sorts exists to the above in the sense that generalized inverses, which are not themselves projectors, have associated with them a pair of projectors.

Theorem 2.6: Given a m×n matrix A and any generalized inverse

$$G = A^+_{\mathcal{S}\mathfrak{F}} + Z, \tag{2.118}$$

1. for G either a {1}- or a {2}-inverse of A

$$GA = \wp_{\mathcal{S},\mathfrak{N}}, \qquad \mathfrak{N} = [A(\mathfrak{F})]^{\perp} \tag{2.119}$$

and

$$AG = \wp_{A(\mathcal{S}),\mathfrak{N}_*}, \qquad \mathfrak{N}_* = \mathfrak{F}^{\perp} \tag{2.120}$$

are full rank projectors;

2. for G a {1.5}-inverse of A

$$GA = \begin{cases} \wp_{\mathcal{S}|\mathfrak{N}}, & \text{if } \mathrm{Col}(ZA) \subset \mathcal{S} \\ \wp^*_{\mathcal{S}|\mathfrak{N}}, & \text{if } \mathrm{Col}(ZA) \not\subset \mathcal{S} \end{cases} \tag{2.121}$$

and

$$AG = \begin{cases} \wp_{A(\mathcal{S})|\mathfrak{N}}, & \text{if } \mathrm{Col}(ZA) \subset \mathfrak{F} \\ \wp^*_{A(\mathcal{S})|\mathfrak{N}}, & \text{if } \mathrm{Col}(ZA) \not\subset \mathfrak{F} \end{cases}, \tag{2.122}$$

where $\mathfrak{N} = \mathrm{Null}(GA)$ and $\mathfrak{N}_* = \mathrm{Null}(AG)$. (2.123)

Proof:

1. Follows immediately since

$$GA = A^+_{\mathcal{S},\mathfrak{F}}A \tag{2.124}$$

and

$$AG = AA^+_{\mathfrak{F},\mathcal{S}}. \tag{2.125}$$

2. (2.121) holds since

$$\mathrm{Rank}(GA) = \mathrm{Dim}(\mathcal{S}) \tag{2.126}$$

if and only if

$$\mathrm{Col}(ZA) \subset \mathcal{S}. \tag{2.127}$$

Likewise (2.122) holds since

$$\mathrm{Rank}(AG) = \mathrm{Dim}(\mathfrak{F}) \tag{2.128}$$

if and only if

$$\mathrm{Col}(Z'A') \subset \mathfrak{F}. \tag{2.129}$$

●

Part 1. of Theorem 2.6 appears in Ben-Israel and Greville [(1974), p. 59].

2.8 {1.5}-Inverses and Solutions to Linear Equations

If Ψ and Γ are disjoint subspaces of $\Re^n$ and $\wp^*_{\Psi|\Gamma}$ is a Rao projector, the solution to the equation

$$\begin{aligned} \underline{x}+\underline{y} &= \underline{z} \\ \underline{x} &\in \Psi \\ \underline{y} &\in \Gamma \end{aligned} \tag{2.130}$$

is given by

$$\begin{aligned} \underline{x} &= \wp^*_{\Psi|\Gamma}\,\underline{z} \\ \underline{y} &= \underline{z} - \underline{x}\,. \end{aligned} \tag{2.131}$$

Although the choice of $\wp^*_{\Psi|\Gamma}$ is not unique, $\underline{x} = \wp^*_{\Psi|\Gamma}\underline{z}$ is always so.

More generally, for an m×n matrix A and subspaces $\Psi \subset \Re^n$ and $\Gamma \subset \Re^m$ such that $A(\Psi)$ and Γ are disjoint, consider the equation

$$\begin{aligned} A\underline{x}+\underline{y} &= \underline{z} \\ \underline{x} &\in \Psi \\ \underline{y} &\in \Gamma. \end{aligned} \tag{2.132}$$

Clearly the solution $(\underline{x},\underline{y})$ is unique if and only if A is a bijective mapping on Ψ. When this is the case, the mapping

$$\underline{z} \mapsto \underline{x} \tag{2.133}$$

is linear and also bijective on $A(\Psi) \oplus \Gamma$. Consequently, there exists an n×m matrix G such that for all $\underline{z} \in A(\Psi) \oplus \Gamma$

$$\begin{aligned} \underline{x} &= G\underline{z} \\ \underline{y} &= (I - AG)\underline{z} \end{aligned} \tag{2.134}$$

is a solution to (2.132). Such a G is a generalized inverse of A. When A is not a bijective mapping on Ψ, the linear mappings yielding solutions to (2.132) are still generalized inverses of A.

<u>Theorem 2.7</u>: Given an m×n matrix A, and subspaces $\Psi \subset \Re^n$ and

$\Gamma \subset \mathfrak{R}^m$ such that $A(\Psi)$ and Γ are disjoint, for any solution to the equation

$$\begin{aligned} A\underline{x}+\underline{y} &= \underline{z} \\ \underline{x} &\in \Psi \\ \underline{y} &\in \Gamma \end{aligned} \tag{2.135}$$

given by a linear mapping

$$\begin{aligned} \underline{x} &= G\underline{z} \in \Psi \\ y &= (I - AG)\underline{z} \in \Gamma. \end{aligned} \tag{2.136}$$

for all $\underline{z} \in A(\Psi) \oplus \Gamma$, it is necessary and sufficient that G be a {1.5}-inverse of A

$$G = A^{+}_{\mathcal{E},\mathcal{F}} + Z \tag{2.137}$$

where

$$A(\Psi) = A(\Psi \cap \mathcal{E}), \tag{2.138}$$

$$\mathcal{F} = \Gamma^{\perp}, \tag{2.139}$$

and

$$Z(\Gamma) = \text{Null}(A). \tag{2.140}$$

Proof:

($\Rightarrow$) By the definition of G, for all $\underline{\psi} \in \Psi$

$$AGA(\underline{\psi}) = A(\underline{\psi}) \tag{2.141}$$

which implies

$$GAGA(\underline{\psi}) = GA(\underline{\psi}) \in \Psi. \tag{2.142}$$

Thus G is a generalized inverse of A of the form (2.137) for some $\mathcal{E}$ and $\mathcal{F}$ where

$$GA(\Psi) \subset \Psi \cap \mathcal{E}. \tag{2.143}$$

Since G is a generalized inverse, it follows from (2.141) that for each $\underline{\psi} \in \Psi$ there exists an $\underline{e} \in \Psi \cap \mathcal{E}$ such that

$$A(\underline{\psi}) = A(\underline{e}) \tag{2.144}$$

or equivalently (2.137). Since G is linear, with $A(\Psi)$ and Γ disjoint, it follows that

$$AG(\Gamma) = \{\underline{0}\}. \tag{2.145}$$

This implies

$$AA^{+}_{\mathcal{E},\mathcal{F}}\underline{\gamma} = -AZ\underline{\gamma}, \text{ for all } \underline{\gamma} \in \Gamma. \tag{2.146}$$

If $\text{Col}(F) = \mathcal{F}$, (2.137) and (2.146) imply

$$F'\underline{\gamma} = -F'AZ\underline{\gamma} = \underline{0}, \text{ for all } \underline{\gamma} \in \Gamma \tag{2.147}$$

or equivalently (2.139). (2.140) now follows from (2.145).

($\Leftarrow$) (2.139) and (2.140) together imply

$$AG(\Gamma) = \{\underline{0}\}. \tag{2.148}$$

If $\underline{\psi} \in \Psi$ let $\underline{e} \in \Gamma \cap \mathcal{S}$ be such that

$$A(\underline{\psi}) = A(\underline{e}), \tag{2.149}$$

then
$$AGA(\underline{\psi}) = AGA(\underline{e}) = A(\underline{e}) = A(\underline{\psi}). \tag{2.150}$$

(2.148) and (2.150) together imply that G given by (2.137) will generate a solution to (2.135).

•

Particular cases of Theorem 2.7 have appeared in the literature. In the special case where $\Gamma = \Re^n$, Col(B) $= \Psi$, and $\Re^m = A(\Gamma) \oplus \Psi$, Rao (1974) identified the linear mapping yielding solutions to (2.135) as a "g-inverse of A constrained by B", $G_{A|B}$, whose general form is

$$G_{A|B} = (C'A)^{-}C' + Z \tag{2.151}$$

where Col(C) $= [\text{Col(B)}]^{\perp}$, AZA = 0 and AZB = 0.

As shown by Bott and Duffin (1953), for a square n×n matrix A and a subspace $\Gamma \subset \Re^n$ such that $\Re^m = A(\Gamma) \oplus \Gamma^{\perp}$, the equation

$$\begin{aligned} A\underline{x} + \underline{y} &= \underline{z} \\ \underline{x} &\in \Gamma \\ \underline{y} &\in \Gamma^{\perp} \end{aligned} \tag{2.152}$$

has a unique solution given by

$$\begin{aligned} \underline{x} &= A_{\Gamma}^{+}\underline{z} \\ \underline{y} &= (I - AA_{\Gamma}^{+})\underline{z} \end{aligned} \tag{2.153}$$

where A_{Γ}^{+} is the Bott-Duffin Inverse of A with respect to Γ.

Given an m×n matrix A and $\underline{z} \in \Re^n$, any $\underline{x}$ minimizing

$$(A\underline{x} - \underline{z})'(A\underline{x} - \underline{z}) \tag{2.154}$$

is known as a <u>least square solution (LSS)</u>. It is well known that finding a LSS to (2.154) is equivalent to finding a solution to (2.134) where $\Gamma = [\text{Col(A)}]^{\perp}$ and $\Psi = \Re^m$. The general form of LSS's is given in the following corollary which may be found in Ben-Israel [(1974), p. 104].

<u>Corollary 2.7.1</u>: The solution to (2.134) where $\Gamma = [\text{Col(A)}]^{\perp}$ and $\Psi = \Re^m$ is given by

$$\begin{aligned} \underline{x} &= G\underline{z} \\ \underline{y} &= (I - AG)\underline{z}, \end{aligned} \tag{2.155}$$

where G is a {1,3}-inverse of A.

Proof: Dim($\mathcal{S}$) = Rank(A), thus $\mathcal{F}$ = Col(A) = A($\mathcal{S}$). By Theorem 2.5, G is a {1,3}-inverse of A.

•

The usual approach for finding LSS's leads to solutions of the form

$$\underline{x} = (A'A)^{-}A'\underline{z}, \tag{2.156}$$

where $(A'A)^{-}$ is any {1}-inverse of A'A. It is readily verified that $(A'A)^{-}A'$ is a {1,2,3}-inverse of A, a null augmented mapping of $[\text{Col}(A)]^{\perp}$ to $\underline{0}$. In contrast, a strict {1,3}-inverse maps at least one vector in $[\text{Col}(A)]^{\perp}$ to some nonnull vector in Null(A). Since the class of {1,3}-inverses includes all {1,2,3}-inverses, this suggests that not all LSS's are given by (2.149). While it is true the nonnull LSS's which lie in Null(A) cannot be generated in the form (2.156), all other LSS's may be so expressed for an appropriate choice of a {1}-inverse of A'A. This implies that it suffices to consider the smaller class of {1,2,3}-inverses rather than the larger class of {1,3}-inverses, if the interest is in generating the nontrivial LSS's which are disjoint from Null(A). In an analagous fashion it is sufficient to consider only the class of {2}-inverses of A to generate the solutions to (2.134) which do not lie in the null space of A.

Corollary 2.2.7: For all $\underline{z} \in A(\Psi) \oplus \Gamma$, all solutions $(\underline{x},\underline{y})$ to (2.135) with $\underline{x}$ not lying in Null(A) may be expressed as

$$\begin{aligned} \underline{x} &= A^{+}_{\mathcal{S},\mathcal{F}}\underline{z} \\ \underline{y} &= \underline{z} - \underline{x} \end{aligned} \tag{2.157}$$

where

$$\mathcal{S} \subset \Psi, \tag{2.158}$$

$$A(\mathcal{S}) = A(\Psi), \tag{2.159}$$

and

$$\mathcal{F} \subset \Gamma^{\perp}. \tag{2.160}$$

Proof: Trivial.

•

The above corollary establishes the notion of {2}-inverses of A as a set of matrices sufficient to generate all linear solutions (2.99). This is an important feature of {2}-inverses which will be further explored in Chapters IV and V.

2.9 Decomposition of {2}-Inverses

Thus far in this chapter, concern has been in characterizing the special role {2}-inverses hold in the broader context of generalized inverses. In this section, attention is focused specifically on {2}-inverses with the aim of outlining several specific results which will be exploited in subsequent chapters.

The starting point for this section is a characterization of {2}-inverses in terms of their full rank factorization which is found in Ben-Israel and Greville [(1974), p. 46].

Lemma 2.8: Given an m×n matrix A and an n×m matrix G written in a full rank factorization $G = EF'$, then

$$G = A^{+}_{\mathcal{E},\mathcal{F}},\ \mathcal{E} = \mathrm{Col}(E),\ \mathcal{F} = \mathrm{Col}(F) \tag{2.161}$$

if and only if

$$F'AE = I. \tag{2.162}$$

Proof:

($\Rightarrow$) $G = A^{+}_{\mathcal{E},\mathcal{F}}$ implies $EF'AEF' = EF'$ which upon premultiplying by $(E'E)^{-1}E'$ and postmultiplying by $F(F'F)^{-1}$ yields (2.162).

($\Leftarrow$) Trivial.

●

Several useful facts concerning {2}-inverses are an immediate consequence of the above lemma. The first concerns the sum of {2}-inverses.

Corollary 2.8.1: If $A^{+}_{\mathcal{E}_i,\mathcal{F}_i}$, $i=1,\ldots,r$ are well defined such that $\underline{f}_i'A\underline{e}_j = 0$ for all $\underline{f}_i \in \mathcal{F}_i$ and $\underline{e}_j \in \mathcal{E}_j$ when $i \neq j$, then

$$A^{+}_{\mathcal{E},\mathcal{F}} = A^{+}_{\mathcal{E}_1,\mathcal{F}_1} + \ldots + A^{+}_{\mathcal{E}_r,\mathcal{F}_r}, \tag{2.163}$$

where $\mathcal{E} = \mathcal{E}_1 \oplus \ldots \oplus \mathcal{E}_r$ and $\mathcal{F} = \mathcal{F}_1 \oplus \ldots \oplus \mathcal{F}_r$.

●

Corollary 2.8.2: Given $A^{+}_{\mathcal{E},\mathcal{F}}$, for any basis $[\underline{e}_1,\cdots,\underline{e}_r]$ of $\mathcal{E}$, there exists a basis $[\underline{f}_1,\cdots,\underline{f}_r]$ of $\mathcal{F}$ such that $\underline{f}_i'A\underline{e}_j = \delta_{ij}$, the Kronecker Delta, and

$$A^{+}_{\mathcal{E},\mathcal{F}} = A^{+}_{\underline{e}_1,\underline{f}_1} + \cdots + A^{+}_{\underline{e}_r,\underline{f}_r}. \tag{2.164}$$

Proof: If $E = [\underline{e}_1|\cdots|\underline{e}_r]$ then there exists a $F = [\underline{f}_1|\cdots|\underline{f}_r]$ such that

$$A^+_{\mathcal{S},\mathcal{F}} = EF' = \underline{e}_1'\underline{f}_1' + \cdots + \underline{e}_r'\underline{f}_r' = A^+_{\underline{e}_1,\underline{f}_1} + \cdots + A^+_{\underline{e}_r,\underline{f}_r} \tag{2.165}$$

●

Conceptually Corollary 2.8.1 may be viewed as follows. Given $A^+_{\mathcal{S},\mathcal{F}}$ and $[\underline{e}_1,\cdots,\underline{e}_r]$, define each $A^+_{\underline{e}_i,\underline{f}_i}$ by

$$A^+_{\underline{e}_i,\underline{f}_i} A\underline{e}_j = \delta_{ij}\,\underline{e}_i. \tag{2.166}$$

Each $A^+_{\underline{e}_i,\underline{f}_i}$ is a {2}-inverse of A and (2.164) trivially follows.

When A is square it is possible to decompose every Bott-Duffin inverse into a sum of one dimensional Bott-Duffin Inverses.

Corollary 2.8.3: Given A symmetric and nonnegative definite, A^+_Γ is nonnegative definite and there exists a basis $[\underline{\gamma}_1,\cdots,\underline{\gamma}_r]$ of Γ such that

$$\delta_{ij} = \underline{\gamma}_i' A \underline{\gamma}_j \tag{2.167}$$

and

$$A^+_\Gamma = A^+_{\underline{\gamma}_1} + \cdots + A^+_{\underline{\gamma}_r}. \tag{2.168}$$

Proof: For G full column rank such that $Col(G) = \Gamma$, $G'AG$ is positive definite and hence also is $(G'AG)^{-1}$. Therefore $A^+_\Gamma = G(G'AG)^{-1}G'$ is nonnegative definite. (2.167) and (2.168) follow immediately.

●

Corollary 2.8.1 outlines the conditions under which a sum of {2}-inverses is itself a {2}-inverse. The following corollary gives a sufficient condition for the difference of two {2}-inverses to be a {2}-inverse.

Corollary 2.8.4: Given spaces $\mathcal{S} \subset \Re^n$ and $\mathcal{F} \subset \Re^m$ both of rank r and subspaces $\mathcal{S}_1 \subset \mathcal{S}$ and $\mathcal{F}_1 \subset \mathcal{F}$ of rank r_1, then

$$A^+_{\mathcal{S},\mathcal{F}} - A^+_{\mathcal{S}_1,\mathcal{F}_1} \text{ is the {2}-inverse } A^+_{\mathcal{S}_0,\mathcal{F}_0} \tag{2.169}$$

where $\mathcal{S}_0 = \mathcal{S} \cap [A'(\mathcal{F}_1)]^\perp$, and $\mathcal{F}_0 = \mathcal{F} \cap [A(\mathcal{S}_1)]^\perp$ satisfy $\mathcal{S} = \mathcal{S}_0 \oplus \mathcal{S}_1$ and $\mathcal{F} = \mathcal{F}_0 \oplus \mathcal{F}_1$.

●

Corollary 2.8.5: If A^+ is the Moore-Penrose Inverse of A then

$$AA^+_{\mathcal{S},\mathcal{F}}A = \left(A^+\right)^+_{A(\mathcal{S}),A'(\mathcal{F})} \tag{2.170}$$

and

$$\left(A^+\right)^+_{\mathcal{S}_0,\mathcal{F}_0} = A - AA^+_{\mathcal{S},\mathcal{F}}A\,, \tag{2.171}$$

where $\mathcal{S}_0 = \mathrm{Col}(A) \cap \left[\mathcal{P}_{\mathcal{C}}(\mathcal{F})\right]^{\perp}$, and $\mathcal{F}_0 = \mathrm{Col}(A') \cap \left[\mathcal{P}_{\mathcal{R}_0}(\mathcal{S})\right]^{\perp}$ for $\mathcal{P}_{\mathcal{R}_0} = A^+A$ and $\mathcal{P}_{\mathcal{C}} = AA^+$ the Euclidean projections into Col(A′) and Col(A), respectively.

Proof: Write $A^+_{\mathcal{S},\mathcal{F}} = EF'$ with $F'AE = I$ as per Lemma 2.8, then $AA^+_{\mathcal{S},\mathcal{F}}A = (AE)(A'F)'$ and $(A'F)'A^+(AE) = I$. (2.170) then follows from Lemma 2.8. From (2.170) and Corollary 2.8.4.

$$A - AA^+_{\mathcal{S},\mathcal{F}}A = \left(A^+\right)^+ - \left(A^+\right)^+_{A(\mathcal{S}),A'(\mathcal{F})} = \left(A^+\right)^+_{\mathcal{S}_0,\mathcal{F}_0}, \tag{2.172}$$

where
$$\mathcal{S}_0 = \mathrm{Col}(A) \cap \left[\left(A^+\right)'A'(\mathcal{F})\right]^{\perp} = \mathrm{Col}(A) \cap \left[AA^+(\mathcal{F})\right]^{\perp} \tag{2.173}$$

and
$$\mathcal{F}_0 = \mathrm{Col}(A') \cap \left[A^+A(\mathcal{F})\right]^{\perp} \tag{2.174}$$

●

(2.171) of the above corollary is a generalization of an equality involving Bott-Duffin Inverses which appears in Ben-Israel and Greville [(1974), p. 90].

$$\left(A^{-1}\right)^+_{\Gamma^{\perp}} = A - AA^+_{\Gamma}A. \tag{2.175}$$

2.10 Spectral Decomposition in Terms of {2}-Inverses

A square matrix A is diagonalizable if it is similar to a diagonal matrix D, that is, if there exists a non-singular matrix P such that $A = P^{-1}DP$. It is well known that A is diagonalizable if and only if A has a spectral decomposition

$$A = \sum_{i=1}^{k} \lambda_i U_i, \tag{2.176}$$

where
$$U_i U_j = \delta_{ij} U_i \tag{2.177}$$

and
$$I = \sum_{i=1}^{k} U_i. \tag{2.178}$$

Matrices U_i are the principal idempotents of A and the scalars λ_i are the eigenvalues of A. A vector $\underline{v}$ is an eigenvector of A if and only $A\underline{v} = \lambda_i\underline{v}$ for some i, in which case $\underline{v} \in \mathrm{Col}(U_i)$. As a consequence of Lemma 2.8, since each U_i is a projection (i.e. a {2}-inverse of the identity), it may be expressed as

$$U_i = E_i F_i' , \tag{2.179}$$

where E_i and F_i are of full column rank r_i such that

$$F_i' E_j = \delta_{ij} I, \tag{2.180}$$

$$F_i' F_j = 0, \text{ for } i \neq j \tag{2.181}$$

and

$$E_i' E_j = 0, \text{ for } i \neq j . \tag{2.182}$$

While not all square matrices have a spectral decomposition, any symmetric is diagonalizable. Nevertheless, any arbitrary matrix has a singular value decomposition

$$A = \sum_{i=1}^{k} \lambda_i E_i F_i' \tag{2.183}$$

where λ_i is the square root of λ_i^2, an eigenvalue of AA' or $A'A$, whose respective spectral decompositions are

$$AA' = \sum_{i=1}^{k} \lambda_i^2 E_i E_i' \tag{2.184}$$

and

$$A'A = \sum_{i=1}^{k} \lambda_i^2 F_i F_i' , \tag{2.185}$$

with E_i, F_i full column rank r_i such that $E_i' E_j = \delta_{ij} I$ and $F_i' F_j = \delta_{ij} I$.

From (2.184) and (2.185) it follows that for all i

$$AA'E_i = \lambda_i^2 E_i \tag{2.186}$$

and

$$A'AF_i = \lambda_i^2 F_i . \tag{2.187}$$

Therefore, it is apparent that $\lambda_i^{-2} A$ is a generalized inverse of A' with respect to $\mathcal{E}_i = \mathrm{Col}(E_i)$and $\mathcal{F}_i = \mathrm{Col}(F_i)$. Theorem 2.3 suggests A may be expressed as

$$A = \sum_{i=1}^{k} \lambda_i^2 (A')^{+}_{\mathcal{E}_i, \mathcal{F}_i} . \tag{2.188}$$

Indeed this is the case! It is easy to verify that

$$(A')^{+}_{\mathcal{E}_i, \mathcal{F}_i} = \lambda_i^{-1} E_i F_i' \tag{2.189}$$

is the {2}-inverse of A with respect to $\mathcal{E}_i$ and $\mathcal{F}_i$. Thus (2.188) follows from

(2.183).

The above implies that the singular value decomposition may be viewed as a partitioning of a matrix into a weighted sum of {2}-inverses of its transpose. More generally, for an $m \times n$ matrix A and an $n \times m$ matrix B, under what conditions is it possible to partition A into a weighted sum of {2}-inverses of B? The following theorem outlines a necessary condition for such a decomposition to hold.

<u>Theorem 2.9</u>: For A, an $m \times n$ matrix, and B, an $n \times m$ matrix, let BA have the spectral decomposition given by

$$BA = \sum_{i=1}^{k} \lambda_i E_i F_i' , \tag{2.190}$$

where E_i, F_i full column rank matrices of rank r_i satisfying $F_i'E_j = \delta_{ij}I$, and $E_i'E_j = 0$, $F_i'F_j = 0$ for $i \neq j$. In addition let $\lambda_0 = 0$ and $\lambda_i \neq 0$ if $i \neq 0$. Then

$$A = (AE_0)F'_0 + \sum_{i=1}^{k} \lambda_i B^{+}_{A(\mathcal{E}_i),\mathcal{F}_i} \tag{2.191}$$

and

$$B = E_0(B'F_0)' + \sum_{i=1}^{k} \lambda_i A^{+}_{\mathcal{E}_i,B'(\mathcal{F}_i)} , \tag{2.192}$$

where $\mathcal{E}_i = \mathrm{Col}(E_i)$ and $\mathcal{F}_i = \mathrm{Col}(F_i)$. Furthermore, for $i \neq j$,

$$\left[A^{+}_{\mathcal{E}_i,B'(\mathcal{F}_i)} \right] A \left[A^{+}_{\mathcal{E}_j,B'(\mathcal{F}_j)} \right] = 0 \tag{2.193}$$

and

$$\left[A^{+}_{A(\mathcal{E}_i),\mathcal{F}_i} \right] A \left[A^{+}_{A(\mathcal{E}_j),\mathcal{F}_j} \right] = 0 . \tag{2.194}$$

<u>Proof</u>: From (2.178)

$$I = E_0F'_0 \quad + \sum_{i=1}^{k} E_iF_i' . \tag{2.195}$$

Thus

$$A = AE_0F'_0 \quad + \sum_{i=1}^{k} AE_iF_i' , \tag{2.196}$$

$$A = (AE_0)F'_0 + \sum_{i=1}^{k} \lambda_i AE_i(F'_i\lambda_i E_i)^{-1}F'_i\ , \tag{2.197}$$

$$A = (AE_0)F'_0 + \sum_{i=1}^{k} \lambda_i AE_i(F'_iBAE_i)^{-1}F'_i\ , \tag{2.198}$$

and

$$A = (AE_0)F'_0 + \sum_{i=1}^{k} \lambda_i B^{+}_{A(\mathcal{E}_i),\mathcal{F}_i}\ . \tag{2.199}$$

Similarly,

$$B = E_0(B'F_0)' + \sum_{i=1}^{k} E_iF'_iB\ , \tag{2.200}$$

$$B = E_0(B'F_0)' + \sum_{i=1}^{k} \lambda_i E_i(F'_i\lambda_i E_i)^{-1}F'_iB\ , \tag{2.201}$$

$$B = E_0(B'F_0)' + \sum_{i=1}^{k} \lambda_i E_i(F'_iBAE_i)^{-1}F'_iB\ , \tag{2.202}$$

and

$$B = E_0(B'F_0)' + \sum_{i=1}^{k} \lambda_i A^{+}_{\mathcal{E}_i,B'(\mathcal{F}_i)}\ . \tag{2.203}$$

●

Expressions (2.191) and (2.192) in the above theorem will be referred to as the Joint {2}-Inverse Decomposition of A and B. As discussed in the introduction to Theorem 2.9, it represents a generalization of both the spectral decomposition of square matrices as well as the singular value decomposition. In Theorem 2.9, the existence of a joint {2}-inverse decomposition depends upon the existence of the spectral decomposition of their product. For arbitrary real matrices A and B, the spectral decomposition of BA is not a given. Nevertheless, under conditions frequently occurring in statistics the spectral decomposition is assured. For example, when A is symmetric and B is nonnegative definite the spectral decomposition always exists. Such a situation arises in the consideration of quadratic forms $\underline{x}'A\underline{x}$ in variables $\underline{x}$ having the variance-covariance matrix

B. In this instance the joint {2}-inverse decomposition of A and B is solely in terms of Bott-Duffin inverses. This is proved in the following corollary.

<u>Corollary 2.9.1</u>: For A n×n symmetric and B n×n nonnegative definite, the spectral decomposition

$$BA = \sum_{i=1}^{k} \lambda_i E_i F_i' \tag{2.204}$$

exists where E_i, F_i are full column rank matrices of rank r_i satisfying $F_i'E_j = \delta_{ij}I$, and $E_i'E_j = 0$, $F_i'F_j = 0$ for $i \neq j$. Under the asumptions $\lambda_0 = 0$, and $\lambda_i \neq 0$ for $i \neq 0$, then

$$A = A_0 + \sum_{i=1}^{k} \lambda_i B^{+}_{\mathcal{F}_i}\,, \tag{2.205}$$

and

$$B = B_0 + \sum_{i=1}^{k} \lambda_i A^{+}_{\mathcal{E}_i}\,. \tag{2.206}$$

where $A_0B = 0$, $B_0A = 0$, $\mathcal{E}_i = \text{Col}(E_i)$ and $\mathcal{F}_i = \text{Col}(F_i)$. Furthermore, for $i \neq j$,

$$A^{+}_{\mathcal{E}_i} A A_{\mathcal{E}_j} = 0 \text{ and } A^{+}_{\mathcal{F}_i} A A_{\mathcal{F}_j} = 0. \tag{2.207}$$

<u>Proof</u>: Write $B = TT'$ where T is full column rank. Since $T'AT$ is symmetric it has a spectral decomposition. Now

$$T'AT\underline{v} = \lambda\underline{v} \text{ implies } BA(T\underline{v}) = \lambda T\underline{v} \tag{2.208}$$

and $$BA\underline{w} = \lambda \quad \text{implies } T'AT(T'A\underline{w}) = \lambda T'A\underline{w}. \tag{2.209}$$

Thus, $T'AT$ and BA have identical eigenvalues occurring with the same multiplicity. Since Rank(BA) = Rank(T'AT), (2.204) follows. For all $i \neq 0$

$$ABF_i = \lambda_i F_i \text{ implies } BA(BF_i) = \lambda_i BF_i \tag{2.210}$$

and $$BAE_i = \lambda_i E_i \text{ implies } AB(AE_i) = \lambda_i AE_i\,. \tag{2.211}$$

Similarly,

$$ABF_0 = 0 \text{ implies } BA(BF_0) = 0 \tag{2.212}$$

and $$BAE_0 = 0 \text{ implies } AB(AE_0) = 0. \tag{2.213}$$

Consequently, for all $i \neq 0$,

$$A^{+}_{\mathcal{E}_i, B'(\mathcal{F}_i)} = A^{+}_{\mathcal{E}_i} \tag{2.214}$$

and $$B^{+}_{A(\mathcal{S}_t),\mathcal{F}_t} = B^{+}_{\mathcal{F}_t} \tag{2.215}$$

Finally,

$$Col(A_0) = A(\mathcal{F}_0) = Col(A) \cap [Col(B)]^{\perp} \tag{2.216}$$

and $$Col(B_0) = A(\mathcal{S}_0) = Col(B) \cap [Col(A)]^{\perp}. \tag{2.217}$$

•

2.11 Computation of {2}-Inverses

Corollary 2.2.1 gives a simple expression for the {2}-inverses of A with respect to the spaces $\mathcal{S}$ and $\mathcal{F}$. However, (2.40) requires full column rank matrices E and F such that $\mathcal{S} = Col(E)$ and $\mathcal{F} = Col(F)$. A less restricted expression for $A^{+}_{\mathcal{S},\mathcal{F}}$ is possible.

For arbitrary matrices E and F, let $(F'AE)^{-}$ be a {1}-inverse of F'AE. Consider matrices of the form

$$E(F'AE)^{-}F' = E(F'AE)^{+}_{1,2}F' + EZF', \tag{2.218}$$

where, in accordance with Theorem 2.4,

$$(F'AE)^{-} = (F'AE)^{+}_{1,2} + Z \tag{2.219}$$

for some {1,2}-inverse $(F'AE)^{+}_{1,2}$ with

$$ZF'AE = 0, \tag{2.220}$$

and $$F'AEZ = 0. \tag{2.221}$$

It is easily verified that

$$A^{+}_{\mathcal{S}_0,\mathcal{F}_0} = E(F'AE)^{+}_{1,2}F', \tag{2.222}$$

where $\mathcal{S}_0 = Col\left[E(F'AE)^{+}_{1,2}\right]$ and $\mathcal{F}_0 = Row\left[(F'AE)^{+}_{1,2}F'\right]$. From Theorem 2.3, $E(F'AE)^{-}F'$, given by (2.218) is a {1.5}-inverse of A and will be a {2}-inverse if and only if

$$EZF' = 0. \tag{2.223}$$

If E and F are matrices such that $\mathcal{S} = Col(E)$ and $\mathcal{F} = Col(F)$ satisfy (2.22), (2.23), (2.24) and (2.25), then either (2.220) or (2.221) implies (2.223). Thus

$$A^{+}_{\mathcal{S},\mathcal{F}} = E(F'AE)^{-}F' \tag{2.224}$$

regardless of the choice of $(F'AE)^{-}$. This result is similar to the one due to Rao and Mitra (1971) quoted earlier in (2.12). The result is also analagous to that appearing in Ben-Israel and Greville [(1974), p. 60]. These results are summarized in the following lemma.

Lemma 2.10: Given matrices E and F, not necessarily full column rank, such that $\mathcal{E} = \mathrm{Col}(E)$ and $\mathcal{F} = \mathrm{Col}(F)$:

1. for an arbitrary {1}-inverse of F′AE

if $$(F'AE)^{-} = (F'AE)^{+}_{1,2} + Z\ , \tag{2.225}$$

then $$G = E(F'AE)^{-}F' \tag{2.226}$$

is a {1.5}-inverse of A with respect to the spaces $\mathcal{E}_0 = \mathrm{Col}\left[E(F'AE)^{+}_{1,2}\right]$ and $\mathcal{F}_0 = \mathrm{Row}\left[(F'AE)^{+}_{1,2}F'\right]$.

2. for any arbitrary {1,2}-inverse $(F'AE)^{+}_{1,2}$

$$A^{+}_{\mathcal{E}_0,\mathcal{F}_0} = E(F'AE)^{+}_{1,2}F, \tag{2.227}$$

where $\mathcal{E}_0 = \mathrm{Col}\left[E(F'AE)^{+}_{1,2}\right]$ and $\mathcal{F}_0 = \mathrm{Row}\left[(F'AE)^{+}_{1,2}F'\right]$;

3. If $A^{+}_{\mathcal{E},\mathcal{F}}$ is well defined, then for any arbitrary {1}-inverse $(F'AE)^{-}$

$$A^{+}_{\mathcal{E},\mathcal{F}} = E(F'AE)^{-}F\ . \tag{2.228}$$

●

The G2SWEEP operator described by Goodnight (1979) provides an efficient way for computing an {1,2}-inverse of an arbitrary matrix. It can be adapted to compute {2}-inverses of A with respect to specified $\mathcal{E}$ and $\mathcal{F}$.

The G2SWEEP may be represented as

$$\begin{bmatrix} A & B \\ C & D \end{bmatrix} \xrightarrow{\text{G2SWEEP}} \begin{bmatrix} A^{+}_{1,2} & A^{+}_{1,2}B \\ -CA^{+}_{1,2} & -CA^{+}_{1,2}B \end{bmatrix} \tag{2.229}$$

where $A^{+}_{1,2}$ is the {1,2}-inverse of the principal minor formed by "pivoting" on the nonzero diagonal elements and zeroing out the rows and columns where there is a zero on the diagonal. From (2.226) is obvious that

$$\begin{bmatrix} F'AE & F' \\ E & 0 \end{bmatrix} \xrightarrow{\text{G2SWEEP}} \begin{bmatrix} (F'AE)^{+}_{1,2} & (F'AE)^{+}_{1,2}F' \\ -E(F'AE)^{+}_{1,2} & -A^{+}_{\mathcal{E}_0,\mathcal{F}_0} \end{bmatrix}, \tag{2.230}$$

where $\mathcal{E}_0 = \mathrm{Col}\left[E(F'AE)^{+}_{1,2}\right]$ and $\mathcal{F}_0 = \mathrm{Row}\left[(F'AE)^{+}_{1,2}F'\right]$. If $A^{+}_{\mathcal{E},\mathcal{F}}$ is well defined and both E and F are choosen such that $\mathrm{Col}(E) = \mathcal{E}$ and $\mathrm{Col}(F) = \mathcal{F}$, then in (2.230), $\mathcal{E}_0 = \mathcal{E}$ and $\mathcal{F}_0 = \mathcal{F}$. This application of the G2SWEEP for computing {2}-inverses will be exploited in Chapter IV.

CHAPTER III
{2}-INVERSES, QUADRATIC FORMS AND SECOND DEGREE POLYNOMIALS

3.0 Introduction

Due to the importance of quadratic forms, for example in the analysis of variance as established by Fisher (1926) and Cochran (1934), the theory of these statistics has been well explored in the statistical literature. Beginning with quadratic forms in normally and independently distributed random variables, Craig (1943) and Hotelling (1944) established the elegant, easily implemented and well known matrix conditions for their independence and χ^2 distribution. Subsequently, in a series of papers by Craig (1947), Matern (1949), Aitken (1950), Carpenter (1950), and Graybill and Marsaglia (1957), these conditions were extended to quadratic forms in correlated normal variables with a positive definite covariance matrix. In this case the conditions still retain the simplicity established by Craig and Hotelling. In turn, these conditions were further extended to quadratic forms in correlated normal variables with only nonnegative definite and possibly singular covariance by Good (1963, 1969, 1970), Khatri (1963, 1977) and Shanbhag (1966, 1968). Unfortunately in this case the conditions become more complicated and lose their previous simplicity.

Quadratic forms, $\underline{x}'A\underline{x}$, are a special case of second degree polynomials, $\underline{x}'A\underline{x}+2\underline{b}'\underline{x}+c$. The extension of the conditions for independence and χ^2 distribution to second degree polynomials appeared in papers by Laha (1956) and Khatri (1962, 1963).

Not all quadratic forms and second degree polynomials necessarily follow a χ^2 distribution. The general distribution of quadratic forms was characterized by Baldessari (1967) when the underlying random variable has a nonsingular covariance. This was later generalized to the case of singular variance by Khatri (1977).

An extensive bibliography on quadratic forms and second degree polynomials may be found in Khatri (1980).

The role of {2}-inverses in quadratic forms and second degree polynomials has not been extensively studied in the literature. Rao (1962) observed that when the covariance matrix is nonsingular, the necessary and sufficient condition for the quadratic form $\underline{x}'A\underline{x}$ to follow a χ^2 distribution is that A be a {2}-inverse of the covariance matrix. However, he incorrectly asserts the same is true when the covariance is singular. As pointed out by Khatri (1963, 1977), when the covariance matrix is singular, the condition of A being a {2}-inverse of the covariance matrix is sufficient, but not necessary, for the quadratic form to follow a χ^2 distribution. Mitra (1968) considered the problem of characterizing the matrices, A, satisfying Khatri's conditions insuring the χ^2 distribution of $\underline{x}'A\underline{x}$.

In this chapter, the role of generalized inverses, especially {2}-inverses, in quadratic forms and second degree polynomials is explored, beginning with the conditions for the χ^2 distribution of quadratic forms. This leads to an expression in terms of {2}-inverses of every χ^2 distributed quadratic form. Consequently, the class of {2}-inverses is a minimally sufficient set of matrices which canonically represents all possible χ^2 distributed quadratic forms. This will be shown to have merit due to the particular properties of {2}-inverses.

For the general case of second degree polynomials, the notion of a canonical representation is expanded to include all possible second degree polynomials whether χ^2 distributed or otherwise. The identification of a class of canonical representatives is important in several ways. First, it permits the easy construction of essentally all second degree polynomials. Secondly, independent polynomials or those following a χ^2 distribution can be readily identified.

There are 9 sections in this chapter following the Introduction, organized as follows:

3.1 The conditions for the independence and χ^2 distribution of quadratic forms and second degree polynomials are reviewed and contrasted for the cases where the variance matrix Σ is singular and nonsingular.

3.2 The conditions on the coefficient matrix A guaranteeing the χ^2

distribution of the associated quadratic forms is further explored. When $|\Sigma| \neq 0$, it is well known that A must be a {2}-inverse of Σ. In contrast, when $|\Sigma| = 0$, the lesser known fact that A must be a particular type of {1.5}-inverse is established in Lemma 3.5.

3.3 In Theorem 3.8, the expression of every χ^2 distributed quadratic form in terms of a Bott-Duffin Inverse is established. This leads to an association between χ^2 distributed quadratic forms and particular vector subspaces.

3.4 When the mean of the underlying normal variable is fixed, different Bott-Duffin Inverses may represent the same quadratic forms almost everywhere. Theorem 3.9 outlines a set of necessary and sufficient conditions for the equality of quadratic forms.

3.5 Lemma 3.10 establishes the class of Bott-Duffin Inverses as the minimal-sufficient set of coefficient matrices for generating all χ^2 distributed quadratic forms for all values of $\underline{\mu}$.

3.6 In Definition 3.1, the notion of Σ-orthogonality is introduced. In Corollary 3.11.1, the statistical independence of χ^2 distributed quadratic forms is shown to be equivalent to the Σ-orthogonality of the associated vector subspaces.

3.7 In this section, the concept of a canonical form of second degree polynomials is introduced as their unique representation almost everywhere. One such canonical form, The First Canonical Form, is established in Theorem 3.15.

3.8 In terms of their First Canonical Form, the conditions for the χ^2 distribution of second degree polynomials are given in Theorem 3.17. As a consequence of the theorem, every χ^2 distributed quadratic form is associated with a unique subspace of Col(Σ).

3.9 The Second Canonical Form of an arbitrary second degree polynomial is established in Theorem 3.18. In terms of this canonical form, the distribution of the polynomial is readily recognized. This representation also establishes each second degree polynomial's association with subspaces of Col(Σ). In Theorem 3.19, the statistical independence of two second degree polynomials is shown to be equivalent to the Σ-orthogonality of the associated spaces.

3.1 χ^2 Distribution And Independence of Quadratic Forms And Second Degree Polynomials

The conditions for the independence and χ^2 distribution of quadratic forms and second degree polynomials serve as the starting point for the present discussion. Some variety exists in these conditions, particularly in those concerning independence when the variance of the underlying normal random variable is singular. A good discussion contrasting these various sets of conditions may be found in Khatri (1978).

Of these various sets of conditions, the set most useful for the present development is that quoted in Searle (1971) and summarized in Theorem 3.1. As is the usual practice, it will be assumed hereafter, without loss of generality, that the coefficient matrix A, of the quadratic form $\underline{x}'A\underline{x}$, is symmetric.

Theorem 3.1: If $\underline{x} \sim \mathcal{N}(\underline{\mu},\Sigma)$, then for

1. Σ nonsingular [Carpenter(1950)],

$$\underline{x}'A\underline{x} \sim \chi^2(r,\tfrac{1}{2}\underline{\mu}'A\underline{\mu}),\ r = \text{Rank}(A)$$

$$\text{if and only if} \tag{3.1}$$

$$A\Sigma A = A;$$

2. Σ singular [Khatri (1963)],

$$x'A\underline{x} \sim \chi^2(r,\tfrac{1}{2}\underline{\mu}'A\underline{\mu}),\ r = \text{trace}(A\Sigma)$$

$$\text{if and only if}$$

$$\Sigma A\Sigma A\Sigma = \Sigma A\Sigma, \tag{3.2}$$

$$\Sigma A\Sigma A\underline{\mu} = \Sigma A\underline{\mu}, \tag{3.3}$$

and

$$\underline{\mu}'A\Sigma A\underline{\mu} = \underline{\mu}'A\underline{\mu}. \tag{3.4}$$

●

Similarly, the condition for the independence of quadratic forms is elegantly simple when the variance is nonsingular but in the case of singular variance it becomes more complex.

Theorem 3.2: If $\underline{x} \sim \mathcal{N}(\underline{\mu},\Sigma)$, then the necessary and sufficient conditions for the independence of $\underline{x}'A\underline{x}$ and $\underline{x}'B\underline{x}$ are for

1. Σ nonsingular [Carpenter(1950)],

$$0 = A\Sigma B; \tag{3.5}$$

2. Σ singular [Khatri (1963)],

$$0 = \Sigma A \Sigma B \Sigma, \tag{3.6}$$

$$\underline{0} = \Sigma A \Sigma B \underline{\mu} = \Sigma B \Sigma A \underline{\mu}, \tag{3.7}$$

and

$$0 = \underline{\mu}' A \Sigma B \underline{\mu}. \tag{3.8}$$

●

The conditions for the independence and χ^2 distribution of second degree polynomials of normal variates were established by Khatri (1963).

Theorem 3.3: Let $\underline{x} \sim N(\underline{\mu},\Sigma)$, $p = \underline{x}'A\underline{x}+2\underline{b}'\underline{x}+c$ and $q = \underline{x}'B\underline{x}+2\underline{m}'\underline{x}+d$, where Σ is possibly singular, then

1. $p \sim \chi^2(r,\frac{1}{2}\delta)$, with $r = \text{Rank}(\Sigma A \Sigma)$, and $\delta = \underline{\mu}'A\underline{\mu}+2\underline{b}'\underline{\mu}+c$, if and only if

$$\Sigma A \Sigma = \Sigma A \Sigma A \Sigma, \tag{3.9}$$

$$\Sigma(A\underline{\mu}+\underline{b}) = \Sigma A \Sigma (A\underline{\mu}+\underline{b}), \tag{3.10}$$

and

$$\underline{\mu}'A\underline{\mu}+2\underline{b}'\underline{\mu}+c = (A\underline{\mu}+\underline{b})'\Sigma(A\underline{\mu}+\underline{b}); \tag{3.11}$$

2. p and q are independent if and only if

$$0 = \Sigma A \Sigma B \Sigma, \tag{3.12}$$

$$0 = \Sigma A \Sigma (B\underline{\mu}+\underline{b}) = \Sigma B \Sigma (A\underline{\mu}+\underline{m}), \tag{3.13}$$

and

$$0 = (A\underline{\mu}+\underline{b})'\Sigma(B\underline{\mu}+\underline{m}). \tag{3.14}$$

●

3.2 Generalized Inverses and Quadratic Forms

The conditions on the coefficient matrix in parts 1. and 2. of Theorem 3.1 are not equivalent. It is trivial to verify that (3.1) always implies (3.2), (3.3) and (3.4) regardless of the rank of Σ or the value of $\underline{\mu}$. Conversely, (3.2) does not necessarily imply (3.1) when Σ is singular. Furthermore, matrices exist satisfying (3.2) but not (3.3) and (3.4) for all $\underline{\mu}$. The following example illustrates these points.

Example 3.1: For

$$\Sigma = \begin{bmatrix} 1 & 0 & 0 \\ 0 & 1 & 0 \\ 0 & 0 & 0 \end{bmatrix} \tag{3.15}$$

and

$$A = \begin{bmatrix} 1 & 0 & 0 \\ 0 & 0 & 1 \\ 0 & 1 & 0 \end{bmatrix}, \tag{3.16}$$

we have

$$\Sigma A \Sigma A \Sigma = \Sigma A \Sigma = \begin{bmatrix} 1 & 0 & 0 \\ 0 & 0 & 0 \\ 0 & 0 & 0 \end{bmatrix} \tag{3.17}$$

and

$$A \neq A \Sigma A = \begin{bmatrix} 1 & 0 & 0 \\ 0 & 0 & 0 \\ 0 & 0 & 1 \end{bmatrix}. \tag{3.18}$$

If $$\underline{\mu}' = \begin{bmatrix} \mu_1 & \mu_2 & \mu_3 \end{bmatrix}, \tag{3.19}$$

then $$\underline{\mu}' A \Sigma A \Sigma = \begin{bmatrix} \mu_1 & 0 & 0 \end{bmatrix}, \tag{3.20}$$

$$\underline{\mu}' A \Sigma = \begin{bmatrix} \mu_1 & \mu_3 & 0 \end{bmatrix}, \tag{3.21}$$

$$\underline{\mu}' A \Sigma A \underline{\mu} = \mu_1^2 + \mu_3^2, \tag{3.22}$$

and $$\underline{\mu}' A \underline{\mu} = \mu_1^2 + 2\mu_2\mu_3. \tag{3.23}$$

Thus A statisfies condition (3.2) but satisfies (3.3) and (3.4) if and only if $\mu_3 = 0$, i.e. if $\underline{\mu} \in \text{Col}(\Sigma)$.

•

Matrices, A, satisfying (3.1) were identified in Chapter II as {2}-inverses of Σ. The characterization of matrices satisfying (3.2) was investigated by Mitra (1968). His result is summarized in the following lemma.

Lemma 3.4: Given B, not necessarily symmetric, then X is a solution of

$$BXBXB = BXB \tag{3.24}$$

if and only if

$$X = H + V, \tag{3.25}$$

where $HBH = H$, $BVB = 0$.

Proof:

($\Rightarrow$) X given by (3.25) trivially satisfies (3.24).

($\Leftarrow$) Let $H = XBXBX$ and $V = X - XBXBX$, then $HBH = H$ and $BVB = 0$.

•

The expression (3.25) for X is similar, but not identical, to that of a generalized inverse given in Chapter II. Although not obvious, it can be easily verified that every matrix of the form (3.25) is a {1.5}-inverse of B.

Lemma 3.5: Given B, not necessarily symmetric, any X of the form (3.25) is a {1.5}-inverse of B,

$$X = B^{+}_{\mathcal{E},\mathcal{F}} + Z, \tag{3.26}$$

where

$$B^{+}_{\mathcal{E},\mathcal{F}} = XBXBX, \tag{3.27}$$

$$\{\underline{0}\} = ZB(\mathcal{E}), \tag{3.28}$$

$$\{\underline{0}\} = Z'B(\mathcal{F}), \tag{3.29}$$

and

$$0 = BZB. \tag{3.30}$$

Proof: Let $B^{+}_{\mathcal{E},\mathcal{F}} = XBXBX$ and $Z = X - XBXBX$.

●

The implication of Lemma 3.5 is that (3.24) is sufficient for X to be a generalized inverse of B. Unfortunately, it is not necessary that all generalized inverses of B satisfy (3.24). This is illustrated in the following example:

Example 3.2: For

$$B = \begin{bmatrix} 2 & 0 & 1 & 0 \\ 0 & 2 & -1 & 0 \\ 1 & -1 & 1 & 0 \\ 0 & 0 & 0 & 0 \end{bmatrix} \tag{3.31}$$

and

$$E' = F' = [0\ 0\ 1\ 0], \tag{3.32}$$

let $\mathcal{E} = \text{Col}(E)$ and $\mathcal{F} = \text{Col}(F)$. Then

$$B^{+}_{\mathcal{E},\mathcal{F}} = \begin{bmatrix} 0 & 0 & 0 & 0 \\ 0 & 0 & 0 & 0 \\ 0 & 0 & 1 & 0 \\ 0 & 0 & 0 & 0 \end{bmatrix}. \tag{3.33}$$

If

$$Z = \begin{bmatrix} 1 & 1 & 0 & 1 \\ 1 & 1 & 0 & 1 \\ 0 & 0 & 0 & 0 \\ 1 & 1 & 0 & 1 \end{bmatrix}, \tag{3.34}$$

let

$$X = B^{+}_{\mathcal{E},\mathcal{F}} + Z = \begin{bmatrix} 1 & 1 & 0 & 1 \\ 1 & 1 & 0 & 1 \\ 0 & 0 & 1 & 0 \\ 1 & 1 & 0 & 1 \end{bmatrix}. \tag{3.35}$$

It follows that

$$XB = X'B' = \begin{bmatrix} 2 & 2 & 0 & 0 \\ 2 & 2 & 0 & 0 \\ 1 & -1 & 1 & 0 \\ 2 & 2 & 0 & 0 \end{bmatrix} \tag{3.36}$$

and
$$XBE = E. \tag{3.37}$$

Thus X is a {1.5}-inverse of B; however,

$$BXB = \begin{bmatrix} 5 & 3 & 1 & 0 \\ 3 & 5 & -1 & 0 \\ 1 & -1 & 1 & 0 \\ 0 & 0 & 0 & 0 \end{bmatrix} \tag{3.38}$$

and

$$BXBXB = \begin{bmatrix} 17 & 15 & 1 & 0 \\ 15 & 17 & -1 & 0 \\ 1 & -1 & 1 & 0 \\ 0 & 0 & 0 & 0 \end{bmatrix}. \tag{3.39}$$

•

3.3 {2}-Inverses and χ^2 Distributed Quadratic Forms

The condition (3.1), that A be a {2}-inverse of Σ, is a natural one for a quadratic form to follow a χ^2 distribution. This is based upon two observations. First from Lemma 2.8, A may be expressed as $A = BB'$ for some full column rank matrix B of rank r, where $I = B'\Sigma B$. Thus

$$\underline{x}'A\underline{x} = \underline{x}'BB'\underline{x} = z'z, \tag{3.40}$$

where
$$\underline{z} \sim \mathcal{N}(B'\underline{\mu}, I). \tag{3.41}$$

Therefore, directly from the χ^2 definition

$$\underline{x}'A\underline{x} \sim \chi^2(r, \tfrac{1}{2}\underline{\mu}'A\underline{\mu}). \tag{3.42}$$

Secondly, consider the following lemma.

Lemma 3.6: For all $\underline{x}$

$$q(\underline{x}) = \underline{x}'A\underline{x} + 2\underline{b}'\underline{x} + c = 0 \tag{3.43}$$

if and only if

$$A = 0,\ \underline{b} = \underline{0} \text{ and } c = 0. \tag{3.44}$$

Proof:

($\Rightarrow$) (3.43) implies

$$\frac{\partial q}{\partial \underline{x}} = 2A\underline{x} + 2\underline{b} = \underline{0} \tag{3.45}$$

and

$$\frac{\partial^2 q}{\partial \underline{x}'\partial \underline{x}} = 2A = 0. \tag{3.46}$$

(3.43), (3.45) and (3.46) imply (3.44).

($\Leftarrow$) Trivial.

•

For an arbitrary matrix A the χ^2 distribution of $\underline{x}'A\underline{x}$ depends upon $\underline{\mu} = E(\underline{x})$ when Σ is singular. The following corollary establishes that $\underline{x}'A\underline{x}$ will follow a χ^2 distribution regardless of the rank of Σ or the value of $\underline{\mu}$, if and only if A is a {2}-inverse of Σ.

Corollary 3.6.1: If $\underline{x} \sim \mathcal{N}(\underline{\mu},\Sigma)$, then for all $\underline{\mu}$

$$\underline{x}'A\underline{x} \sim \chi^2(r, \tfrac{1}{2}\underline{\mu}'A\underline{\mu}),\ r = \text{trace}(A\Sigma)$$

if and only if

$$A\Sigma A = A. \tag{3.47}$$

Proof:

($\Rightarrow$) $\underline{\mu}'A\Sigma A\underline{\mu} = \underline{\mu}'A\underline{\mu}$ for all $\underline{\mu}$ implies (3.47).

($\Leftarrow$) Trivial.

•

This leads to a problem of determining if all χ^2 distributed quadratic forms can be represented in terms of {2}-inverses of Σ and if this representation is unique. This problem was considered in passing by Mitra (1968) in the paper previously quoted. There he points out that if $\underline{x}'A\underline{x}$ follows a χ^2 distribution, then there exists a nonnegative definite matrix, which happens to be a {2}-inverse of Σ, such that $\underline{x}'A\underline{x} = \underline{x}'\Sigma_{\Gamma}^{+}\underline{x}$ a.e.. The proof of this assertion is a simple consequence of Corollary 3.6.1 and the following lemma.

Lemma 3.7: If $\underline{x} \sim \mathcal{N}(\underline{\mu},\Sigma)$ then

$$\underline{x}'A\underline{x}+2\underline{b}'\underline{x}+c = 0 \text{ a.e.}$$

if and only if

$$\underline{0} = \Sigma A \Sigma, \tag{3.48}$$

$$0 = \Sigma(\underline{b}+A\underline{\mu}), \tag{3.49}$$

and

$$0 = \underline{\mu}'A\underline{\mu}+2\underline{b}'\underline{\mu}+c\ . \tag{3.50}$$

Proof: Writing $\Sigma = BB'$ where B is full column, there exists $\underline{z} \sim \mathcal{N}(\underline{0},I)$ such that

$$\underline{x} = B\underline{z}+\underline{\mu} \text{ a.e.} \tag{3.51}$$

Thus from Lemma 3.6

$$\underline{x}'A\underline{x}+2\underline{b}'\underline{x}+c = \begin{bmatrix} \underline{z}'B'AB\underline{z}+2(A\underline{\mu}+\underline{b})'B\underline{z}+ \\ \underline{\mu}'A\underline{\mu}+2\underline{b}'\underline{\mu}+c \end{bmatrix} = 0 \text{ a.e.} \tag{3.52}$$

if an only if

$$0 = B'AB, \tag{3.53}$$

$$\underline{0} = B'(\underline{b}+A\underline{\mu}), \tag{3.54}$$

and

$$0 = \underline{\mu}'A\underline{\mu}+2\underline{b}'\underline{\mu}+c\ . \tag{3.55}$$

Since (3.53) is equivalent to (3.48) and (3.54) is equivalent to (3.49) the lemma follows immediately.

●

This leads to the following restatement of Mitra's result.

Theorem 3.8: If $\underline{x} \sim \mathcal{N}(\underline{\mu},\Sigma)$ then

$$\underline{x}'A\underline{x} \sim \chi^2(r,\tfrac{1}{2}\underline{\mu}'A\underline{\mu}) \tag{3.56}$$

if and only if

$$\underline{x}'A\underline{x} = \underline{x}'\Sigma_\Gamma^+\underline{x} \text{ a.e.,} \tag{3.57}$$

for the r-dimensional subspace $\Gamma = \text{Col}(A\Sigma A\Sigma A)$.

Proof:

($\Rightarrow$) From Lemma 3.5 the coefficient matrix may be expressed as

$$A = \Sigma_\Gamma^+ + Z, \tag{3.58}$$

where

$$0 = Z\Sigma\Sigma_\Gamma^+ \tag{3.59}$$

and

$$0 = \Sigma Z \Sigma. \tag{3.60}$$

As a consequence of (3.59) and (3.60) it is easy to verify, following some algebra, that (3.2) and (3.3) hold if and only if

$$\underline{0} = \Sigma Z \underline{\mu}, \tag{3.61}$$

and

$$0 = \underline{\mu}' Z \underline{\mu}. \tag{3.62}$$

Thus, by Lemma 3.6

$$\underline{x}' Z \underline{x} = 0 \text{ a.e.} \tag{3.63}$$

or equivalently

$$\underline{x}' A \underline{x} = \underline{x}' \Sigma_{\Gamma}^{+} \underline{x} \text{ a.e.} \tag{3.64}$$

($\Leftarrow$) Trivial.

●

3.4 On The Uniqueness of the {2}-Inverse Representation of χ^2 Distributed Quadratic Forms

Theorem 3.8 indicates that χ^2 distributed quadratic forms are associated with {2}-inverses and, in turn, with subspaces Γ where $\Gamma \cap \text{Null}(\Sigma) = \{\underline{0}\}$. Is this association between subspaces and quadratic forms unique? Unfortuately the answer is no, as illustrated by the following example.

Example 3.3: If $\underline{x} \sim \mathcal{N}(\underline{\mu}, \Sigma)$, where

$$\Sigma = \begin{bmatrix} 5 & 4 & 4 & 5 \\ 4 & 5 & 5 & 4 \\ 4 & 5 & 5 & 4 \\ 5 & 4 & 4 & 5 \end{bmatrix} \tag{3.65}$$

then

$$x_1 - \mu_1 = x_4 - \mu_4 \text{ a.e.} \tag{3.66}$$

and

$$x_2 - \mu_2 = x_3 - \mu_3 \text{ a.e.} \tag{3.67}$$

For

$$\underline{\lambda}' = \begin{bmatrix} 0 & 1 & 1 & 0 \end{bmatrix} \tag{3.68}$$

and

$$\underline{\omega}' = \begin{bmatrix} 1 & 2 & 0 & -1 \end{bmatrix}, \tag{3.69}$$

the corresponding {2}-inverses of Σ are

$$\Sigma_{\underline{\lambda}}^{+} = \frac{1}{20} \begin{bmatrix} 0 & 0 & 0 & 0 \\ 0 & 1 & 1 & 0 \\ 0 & 1 & 1 & 0 \\ 0 & 0 & 0 & 0 \end{bmatrix} \tag{3.70}$$

and

$$\Sigma_{\underline{\omega}}^{+} = \frac{1}{20}\begin{bmatrix} 1 & 2 & 0 & -1 \\ 2 & 4 & 0 & -2 \\ 0 & 0 & 0 & 0 \\ -1 & -2 & 0 & 1 \end{bmatrix}. \tag{3.71}$$

2Using (3.66) and (3.67) the quadratic forms associated with these {2}-inverses may be written as

$$\underline{x}'\Sigma_{\underline{\lambda}}^{+}\underline{x} = \tfrac{1}{20}\left(2x_2 + \mu_3 - \mu_2\right)^2 \tag{3.72}$$

and

$$\underline{x}'\Sigma_{\underline{\omega}}^{+}\underline{x} = \tfrac{1}{20}\left(2x_2 + \mu_1 - \mu_4\right)^2. \tag{3.73}$$

Consequently,

$$\underline{x}'\Sigma_{\underline{\lambda}}^{+}\underline{x} = \underline{x}\Sigma_{\underline{\omega}}^{+}\underline{x} \text{ a.e.}$$

$$\text{if and only if} \tag{3.74}$$

$$\mu_1 + \mu_2 = \mu_3 + \mu_4 .$$

●

The following theorem establishes the necessary and sufficient conditions for the equality almost everywhere of χ^2 distributed quadratic forms when $\underline{\mu}$ is fixed.

Theorem 3.9: If $\underline{x} \sim \mathcal{N}(\underline{\mu},\Sigma)$, then

$$\underline{x}'\Sigma_{\Gamma}^{+}\underline{x} = \underline{x}'\Sigma_{\Psi}^{+}\underline{x} \text{ a.e.}$$

$$\text{if and only if}$$

$$\Sigma(\Gamma) = \Sigma(\Psi) = \mathcal{S} \subset \text{Col}(\Sigma) \tag{3.75}$$

and

$$\mathcal{P}_{\mathcal{S}|\Gamma^{\perp}}\underline{\mu} = \mathcal{P}_{\mathcal{S}|\Psi^{\perp}}\underline{\mu} . \tag{3.76}$$

Proof: By Lemma 3.5

$$\underline{x}'\Sigma_{\Gamma}^{+}\underline{x} = \underline{x}'\Sigma_{\Psi}^{+}\underline{x} \text{ a.e.}$$

$$\text{if and only if}$$

$$\Sigma\Sigma_{\Gamma}^{+}\Sigma = \Sigma\Sigma_{\Psi}^{+}\Sigma, \tag{3.77}$$

$$\Sigma\Sigma_{\Gamma}^{+}\underline{\mu} = \Sigma\Sigma_{\Psi}^{+}\underline{\mu}, \tag{3.78}$$

and

$$\underline{\mu}'\Sigma_{\Gamma}^{+}\underline{\mu} = \underline{\mu}'\Sigma_{\Psi}^{+}\underline{\mu}. \tag{3.79}$$

Since

$$\Sigma\Sigma_{\Gamma}^{+}\Sigma = \left(\Sigma^{+}\right)_{\Sigma(\Gamma)}^{+} \tag{3.80}$$

and

$$\Sigma\Sigma_{\Psi}^{+}\Sigma = \left(\Sigma^{+}\right)_{\Sigma(\Psi)}^{+}, \tag{3.81}$$

(3.77) implies $$\Sigma(\Gamma) = \Sigma(\Psi). \tag{3.82}$$

Condition (3.78) implies (3.79). Since (3.75) holds,

$$\Sigma\Sigma_{\Gamma}^{+} = \wp_{\mathcal{S}|\Gamma^{\perp}} \tag{3.83}$$

and

$$\Sigma\Sigma_{\Psi}^{+} = \wp_{\mathcal{S}|\Psi^{\perp}} \tag{3.84}$$

together with (3.78) imply (3.76).

•

3.5 A Minimal Sufficient Set of Coefficient Matrices For All χ^2 Distributed Quadratic Forms

At first glance Theorem 3.9 appears to suggest that a proper subset of the {2}-inverses of Σ is sufficient to generate all χ^2 distributed quadratic forms. While this is true when $\underline{\mu}$ is restricted to some subspace, it is necessary to consider the entire class of {2}-inverses of Σ in order to generate all χ^2 distributed quadratic forms when $\underline{\mu}$ is unrestricted. This follows from (3.4) since

$$\underline{x}'\Sigma_{\Gamma}^{+}\underline{x} = \underline{x}'\Sigma_{\Psi}^{+}\underline{x}, \text{ a.e. for all } \underline{\mu} \tag{3.85}$$

implies

$$\underline{\mu}'\Sigma_{\Gamma}^{+}\underline{\mu} = \underline{\mu}'\Sigma_{\Psi}^{+}\underline{\mu}, \text{ for all } \underline{\mu} \tag{3.86}$$

or equivalently

$$\Sigma_{\Gamma}^{+} = \Sigma_{\Psi}^{+}. \tag{3.87}$$

The converse is immediate. In this sense the class of {2}-inverses of Σ is the <u>minimal-sufficient</u> set of coefficient matrices for generating all χ^2 distributed quadratic forms for all values of $\underline{\mu}$. This is summarized in the following lemma.

<u>Lemma 3.10</u>: For $\underline{x} \sim \mathcal{N}(\underline{\mu},\Sigma)$, then for all $\underline{\mu}$

$$\underline{x}'\Sigma_{\Gamma}^{+}\underline{x} = \underline{x}'\Sigma_{\Psi}^{+}\underline{x}, \text{ a.e.} \tag{3.88}$$

if and only if

$$\Gamma = \Psi. \tag{3.89}$$

•

If one is willing to restrict the space to which $\underline{\mu}$ belongs then a proper subset of the {2}-inverses of Σ may be sufficient to generate all χ^2 distributed quadratic forms. Such a situation arises naturally in the weighted least squares analysis of categorical data which has been popularized by Grizzle, Starmer and Koch (1969). In the GSK approach, inference is based upon the sample multinomial cell probabilities, $\hat{\underline{\Pi}}$, whose covariance matrix, V, is singular and whose expectation satisfies $E(\hat{\underline{\Pi}}) \in Col(V)$.

In such a circumstance where $\underline{\mu}$ is restricted to Col(Σ), a minimal sufficient set of matrices generating all χ^2 distributed quadratic forms is given by the set of {2,3,4}-inverses of Σ, i.e. the subset of {2}-inverses where $\Gamma \subset Col(\Sigma)$. This is a consequence of the following corollary to Theorem 3.9.

Corollary 3.9.1: For $\underline{x} \sim \mathcal{N}(\underline{\mu},\Sigma)$

1. if $\Pi = \wp_\Sigma(\Psi) \subset Col(\Sigma)$, then for all $\underline{\mu} \in Col(\Sigma)$

$$\underline{x}'\Sigma^+_\Pi\underline{x} = \underline{x}'\Sigma^+_\Psi\underline{x} \quad \text{a.e.} \tag{3.90}$$

where $\wp_\Sigma$ is the Euclidian projection into Col(Σ);

2. If Γ and Ψ are subspaces of Col(Σ) then, for all $\underline{\mu} \in Col(\Sigma)$

$$\underline{x}'\Sigma^+_\Gamma\underline{x} = \underline{x}'\Sigma^+_\Psi\underline{x} \quad \text{a.e.} \tag{3.91}$$

if and only if

$$\Gamma = \Psi\,. \tag{3.92}$$

Proof:

1. For all $\underline{\mu} \in Col(\Sigma)$

$$\Sigma\Sigma^+_\Pi\underline{\mu} = \Sigma\Sigma^+_\Psi\underline{\mu}\,. \tag{3.93}$$

(⇒) We will prove the converse. If $\Gamma \neq \Psi$ then there exists $\underline{\mu} \in Col(\Sigma)$ which is orthogonal to Γ but not to Ψ. Thus

$$0 = \Sigma\Sigma^+_\Gamma\underline{\mu} \neq \Sigma\Sigma^+_\Psi\underline{\mu} \tag{3.94}$$

implying

$$\underline{x}'\Sigma^+_\Gamma\underline{x} \neq \underline{x}'\Sigma^+_\Psi\underline{x}. \tag{3.95}$$

(⇐) Trivial.

•

3.6 Independence of χ^2 Distributed Quadratic Forms

As observed for the conditions guaranteeing the χ^2 distribution of quadratic forms, the conditions for the independence of quadratic forms when Σ is singular and nonsingular are not equivalent. While (3.5) implies (3.6), (3.7) and (3.8) regardless of the rank of Σ and the value of $\underline{\mu}$, (3.6) does not imply (3.5). Furthermore, an A satisfying (3.6) may satisfy (3.7) and (3.8) for only selected values of $\underline{\mu}$. The following continuation of Example 3.1 illustrates these points.

Example 3.1 (continued): For Σ given by (3.15), A given by (3.16) and B given by

$$B = \begin{bmatrix} 0 & 0 & 1 \\ 0 & 0 & 1 \\ 1 & 1 & 1 \end{bmatrix} \tag{3.96}$$

then

$$A\Sigma B = \begin{bmatrix} 0 & 0 & 1 \\ 0 & 0 & 0 \\ 0 & 0 & 1 \end{bmatrix} \tag{3.97}$$

and

$$\Sigma A\Sigma B\Sigma = \begin{bmatrix} 0 & 0 & 1 \\ 0 & 0 & 0 \\ 0 & 0 & 1 \end{bmatrix}. \tag{3.98}$$

Furthermore,

$$\underline{\mu}'A\Sigma B\Sigma = \begin{bmatrix} 0 & 0 & 0 \end{bmatrix}, \tag{3.99}$$

$$\underline{\mu}'B\Sigma A\Sigma = \begin{bmatrix} \mu_3 & 0 & 0 \end{bmatrix}, \tag{3.100}$$

and

$$\underline{\mu}'B\Sigma A\underline{\mu} = \mu_3(\mu_1 + \mu_3). \tag{3.101}$$

Consequently, conditions (3.7) and (3.8) hold if and only if $\mu_3 = 0$, i.e. if $\underline{\mu} \in \text{Col}(\Sigma)$.

●

Condition (3.5) is a natural one for the independence of two quadratic forms in the sense that (3.5) is necessary and sufficient for the independence of two quadratic forms for all choices of $\underline{\mu}$.

Lemma 3.11: If $\underline{x} \sim \mathcal{N}(\underline{\mu},\Sigma)$ then for all $\underline{\mu}$

$\underline{x}'A\underline{x}$ and $\underline{x}'B\underline{x}$ are independent

if and only if (3.102)

$$0 = A\Sigma B. \tag{3.103}$$

Proof:

($\Rightarrow$) For all $\underline{\mu}$

$$0 = \underline{\mu}'A\Sigma B\underline{\mu} \tag{3.104}$$

thus

$$0 = A\Sigma B. \tag{3.105}$$

($\Leftarrow$) Trivial.

●

The circumstances under which $A\Sigma B = 0$ is necessary and sufficient for the independence of $\underline{x}'A\underline{x}$ and $\underline{x}'B\underline{x}$ has been explored in the literature. One such condition is when A and B are positive definite which follows from the matrix result established by Shanbhag (1966).

Lemma 3.12: If A, B and Σ are nonnegative definite, then

$$0 = \Sigma A\Sigma B\Sigma \tag{3.106}$$

if and only if

$$0 = A\Sigma B. \tag{3.107}$$

Proof: Write $A = UU'$ and $B = VV'$ where U and V are full column rank.

($\Rightarrow$)

$$(3.106) \Rightarrow 0 = \Sigma A\Sigma B\Sigma A\Sigma = (V'\Sigma A\Sigma)'(V'\Sigma A\Sigma) \tag{3.108}$$

$$\Rightarrow 0 = V'\Sigma A\Sigma \tag{3.109}$$

$$\Rightarrow 0 = V'\Sigma A\Sigma V = V'\Sigma UU'\Sigma V = (U'\Sigma V)'(U'\Sigma V) \tag{3.110}$$

$$\Rightarrow 0 = U'\Sigma V \tag{3.111}$$

$$\Rightarrow 0 = A\Sigma B. \tag{3.112}$$

($\Leftarrow$) Trivial.

●

Consequently, when A and B are positive definite, condition (3.5) alone is the necessary and sufficient condition for the independence of the associated quadratic forms. In particular, this is true for χ^2 distributed quadratic forms represented in terms of {2}-inverses of Σ since every {2}-inverse of Σ is nonnegative definite.

Corollary 3.12.1: For $\underline{x} \sim \mathcal{N}(\underline{\mu},\Sigma)$

$\underline{x}'\Sigma_\Gamma^+\underline{x}$ and $\underline{x}'\Sigma_\Psi^+\underline{x}$ are independent

if and only if

$$\underline{\gamma}'\Sigma\underline{\psi} = 0, \text{ for all } \underline{\gamma} \in \Gamma \text{ and } \underline{\psi} \in \Psi. \tag{3.113}$$

●

Condition (3.113) on subspaces Γ and Ψ appeared in Chapter II and will appear repeatedly in the remainder of this monograph. To formalize this condition the following definition is made.

Definition 3.1: Given a nonnegative definite matrix Σ, subspaces Γ and Ψ are said to be Σ-orthogonal if and only if

$$\underline{\gamma}'\Sigma\underline{\psi} = 0, \text{ for all } \underline{\gamma} \in \Gamma \text{ and } \underline{\psi} \in \Psi. \tag{3.114}$$

●

As noted earlier in Section 3.3, the quadratic forms in the above corollary may be uniquely represented as

$$\underline{x}'\Sigma_{\Gamma}^{+}\underline{x} = \underline{x}'G'G\underline{x} \tag{3.115}$$

and

$$\underline{x}'\Sigma_{\Psi}^{+}\underline{x} = \underline{x}'J'J\underline{x}\ , \tag{3.116}$$

where G and J are full column rank such that

$$G'\Sigma G = I \text{ and } J'\Sigma J = I. \tag{3.117}$$

Since the above corollary implies $G'\Sigma J = 0$, the independence of the quadratic forms is equivalent to the independence of the underlying normal variates $J'\underline{x}$ and $G'\underline{x}$.

Corollary 3.12.2: Let $\underline{x} \sim \mathcal{N}(\underline{\mu},\Sigma)$,

$$p = \underline{x}'\Sigma_{\Gamma}^{+}\underline{x} = \underline{x}'G'G\underline{x}\ , \tag{3.118}$$

$$q = \underline{x}'\Sigma_{\Psi}^{+}\underline{x} = \underline{x}'J'J\underline{x} \tag{3.119}$$

and

$$r = H'\underline{x}\ , \tag{3.120}$$

where G and J are full column rank with $G'\Sigma G = I$ and $J'\Sigma J = I$. Then

1. p and q are independent if and only if $G'\Sigma J = 0$; (3.121)
2. p and r are independent if and only if $G'\Sigma H = 0$. (3.122)

●

3.7 A Canonical Representation of Second Degree Polynomials

Up to this point in the chapter attention has been restricted to χ^2 distributed quadratic forms. Due to the necessary and sufficient conditions on the coefficient matrix, the set of {2}-inverse was established, in a natural fashion, as a minimal sufficient set of coefficient matrices which canonically represents all χ^2 distributed quadratic forms. Unfortunately, no such conditions exist, a priori, on the coefficient matrix of an arbitrary quadratic form on which to establish a canonical representation. Nevertheless, it is possible to identify a unique canonical representative. This requires considering the more general problem of finding a canonical representative of an arbitrary second degree polynomial. As for χ^2 distributed quadratic forms, establishing a canonical representative for an arbitrary second degree polynomial follows directly as a consequence of Lemma 3.7.

Lemma 3.13: If $\underline{x} \sim \mathcal{N}(\underline{\mu},\Sigma)$, then every second degree polynomial

$$q = \underline{x}'A\underline{x} + 2\underline{b}'\underline{x} + c \tag{3.123}$$

has a representation as

$$q = \underline{x}'B\underline{x} + 2m'x + d, \tag{3.124}$$

where $\mathrm{Col}(B) \subset \mathrm{Col}(\Sigma)$ and $\underline{m} \in \mathrm{Col}(\Sigma)$.

Proof: If Σ^+ is the Moore-Penrose inverse of Σ, let

$$B = \Sigma^+\Sigma A \Sigma\Sigma^+, \tag{3.125}$$

$$\underline{m} = \Sigma^+\Sigma(\underline{b} + A\underline{\mu}_0), \tag{3.126}$$

and

$$d = \underline{\mu}_0' A \underline{\mu}_0 + 2\underline{b}'\underline{\mu}_0 + c, \tag{3.127}$$

where $\underline{\mu} = \underline{\mu}_0 \oplus \underline{\mu}_1$, $\underline{\mu}_0 \in \mathrm{Null}(\Sigma)$ and $\underline{\mu}_1 \in \mathrm{Col}(\Sigma)$. If q is expressed as

$$q = (\underline{x} - \underline{\mu})'A(\underline{x} - \underline{\mu}) + 2(A\underline{\mu} + \underline{b})'(\underline{x} - \underline{\mu}) + \underline{\mu}'A\underline{\mu} + 2\underline{b}'\underline{\mu} + c, \tag{3.128}$$

then by Lemma 3.7 the following is true almost everywhere:

$$q = (\underline{x} - \underline{\mu})'B(\underline{x} - \underline{\mu}) + 2(A\underline{\mu} + \underline{b})'\Sigma\Sigma^+(\underline{x} - \underline{\mu}) + \underline{\mu}'A\underline{\mu} + 2\underline{b}'\underline{\mu} + c \tag{2.129}$$

$$\begin{aligned} = {} & \underline{x}'B\underline{x} - 2\underline{\mu}'B\underline{x} + 2(A\underline{\mu} + \underline{b})'\Sigma\Sigma^+\underline{x} - 2(A\underline{\mu} + \underline{b})'\Sigma\Sigma^+\underline{\mu} \\ & + \underline{\mu}'B\underline{\mu} + \underline{\mu}'A\underline{\mu} + 2\underline{b}'\underline{\mu} + c \end{aligned} \tag{3.130}$$

$$\begin{aligned} = {} & \underline{x}'B\underline{x} + 2(A\underline{\mu} + \underline{b} - A\underline{\mu}_1)'\Sigma\Sigma^+\underline{x} - 2(A\underline{\mu} + \underline{b})'\underline{\mu}_1 \\ & + \underline{\mu}_1'A\underline{\mu}_1 + \underline{\mu}'A\underline{\mu} + 2\underline{b}'\underline{\mu} + c \end{aligned} \tag{3.131}$$

$$= \underline{x}'B\underline{x} + 2(A\underline{\mu}_0 + \underline{b})'\Sigma\Sigma^+\underline{x} + \underline{\mu}_0'A\underline{\mu}_0 + 2\underline{b}'\underline{\mu}_0 + c\ . \tag{3.132}$$

●

The following definition is motivated by Lemma 3.13.

Definition 3.2: Every second degree polynomial written as

$$\underline{x}'B\underline{x}+2\underline{m}'\underline{x}+d, \tag{3.133}$$

where Col(B) $\subset$ Col(Σ) and $\underline{m} \in$ Col(Σ), is said to be in First Canonical Form.

•

Recall that for χ^2 distributed quadratic forms, their representation in terms of {2}-inverses is not unique in that different {2}-inverses may yield equivalent quadratic forms. In contrast, the following lemma establishes that every second degree polynomial in First Canonical Form is unique.

Lemma 3.14: For $\underline{x} \sim \mathcal{N}(\underline{\mu},\Sigma)$, if

$$q_1 = \underline{x}'B_1\underline{x}+2\underline{m}_1'\underline{x}+\underline{d}_1 \tag{3.134}$$

and

$$q_2 = \underline{x}'B_2\underline{x}+2\underline{m}_2'\underline{x}+\underline{d}_2 \tag{3.135}$$

are in First Canonical Form, then

$$q_1 = q_2 \text{ a.e.}$$
$$\text{if and only if} \tag{3.136}$$
$$B_1 = B_2, \underline{m}_1 = \underline{m}_2 \text{ and } d_1 = d_2 .$$

Proof:

($\Rightarrow$) Since Col(B_i) $\subset$ Col(Σ) and $\underline{m}_i \in$ Col(Σ), for i = 1, 2, it follow that

$$\Sigma B_1\Sigma = \Sigma B_2\Sigma$$
$$\text{if and only if} \tag{3.137}$$
$$B_1 = B_2.$$

Given (3.137)

$$\Sigma(\underline{m}_1+B_1\underline{\mu}) = \Sigma(\underline{m}_2+B_2\underline{\mu})$$
$$\text{if and only if} \tag{3.138}$$
$$\underline{m}_1 = \underline{m}_2.$$

Given (3.137) and (3.138)

$$\underline{\mu}'B_1\underline{\mu}+2\underline{m}_1'\underline{\mu}+d_1 = \underline{\mu}'B_2\underline{\mu}+2\underline{m}_2'\underline{\mu}+d_2$$
$$\text{if and only if} \tag{3.139}$$
$$d_1 = d_2.$$

($\Leftarrow$) Trivial.

•

Lemma 3.13 and Lemma 3.14 together imply the uniqueness of the First Canonical Representation of second degree polynomials. This is formalized in the following theorem.

Theorem 3.15: Every second degree polynomial in $\underline{x} \sim \mathcal{N}(\underline{\mu},\Sigma)$ has a unique representation of the form

$$\underline{x}'B\underline{x}+2\underline{m}'\underline{x}+d, \tag{3.140}$$

where $Col(B) \subset Col(\Sigma)$ and $\underline{m} \in Col(\Sigma)$.

•

3.8 χ^2 Distributed Second Degree Polynomials

As they did for χ^2 distributed quadratic forms, {2}-inverses have a key role in establishing a canonical form for χ^2 distributed second degree polynomials. This is done in two stages beginning with second degree polynomials expressed in First Canonical Form.

Lemma 3.16: For $\underline{x} \sim \mathcal{N}(\underline{\mu},\Sigma)$, let

$$q = \underline{x}'B\underline{x}+2\underline{m}'\underline{x}+d$$

be in First Canonical Form, then $q \sim \chi^2(r,\frac{1}{2}\delta)$ if and only if $B = \Sigma_\Gamma^+$, $\underline{m} \in \Gamma$, and $d = \underline{m}'\Sigma\underline{m}$ for an r-dimensional subspace $\Gamma = Col(B) \subset Col(\Sigma)$. Under the conditions of the lemma, q may be uniquely expressed as

$$q = (\underline{x}+\Sigma\underline{m})'\Sigma_\Gamma^+(\underline{x}+\Sigma\underline{m}). \tag{3.141}$$

Proof:

($\Rightarrow$) From Theorem 3.3

$$q \sim \chi^2$$

if and only if

$$\Sigma B\Sigma B\Sigma = \Sigma B\Sigma, \tag{3.142}$$

$$\Sigma B\Sigma(B\underline{\mu}+\underline{m}) = \Sigma(B\underline{\mu}+\underline{m}), \tag{3.143}$$

and

$$\underline{\mu}'B\underline{\mu}+2\underline{\mu}'\underline{m}+d = (B\underline{\mu}+\underline{m})'\Sigma(B\underline{\mu}+\underline{m}). \tag{3.144}$$

Since $\Gamma \subset Col(\Sigma)$, (3.142) holds if and only if $B\Sigma B = B$. In turn, since $\underline{m} \in \Sigma$ and B is a {2}-inverse of Σ, (3.143) holds if and only if $B\Sigma\underline{m} = \underline{m} \in Col(B) = \Gamma$. Finally,

$$\begin{aligned}(B\underline{\mu}+\underline{m})'\Sigma(B\underline{\mu}+\underline{m}) &= \underline{\mu}'B\Sigma B\underline{\mu}+2\underline{\mu}'B\Sigma\underline{m}+\underline{m}'\Sigma\underline{m} \\ &= \underline{\mu}'B\underline{\mu}+2\underline{\mu}'\underline{m}+\underline{m}'\Sigma\underline{m} .\end{aligned} \tag{3.145}$$

Thus (3.144) holds if and only if $d = \underline{m}'\Sigma\underline{m}$. The above leads to the following equality

$$q = \underline{x}'\Sigma_\Gamma^+\underline{x}+2\underline{m}'\Sigma\Sigma_\Gamma^+\underline{x}+\underline{m}'\Sigma\Sigma_\Gamma^+\Sigma\underline{m} = (\underline{x}+\Sigma\underline{m})'\Sigma_\Gamma^+ (\underline{x}+\Sigma\underline{m}). \tag{3.146}$$

($\Leftarrow$) Trivially follows from (3.146).

•

The canonical expression for any χ^2 distributed second degree polynomial is given in the following:

Theorem 3.17: For $\underline{x} \sim \mathcal{N}(\underline{\mu},\Sigma)$, if Σ^+ is the Moore-Penrose Inverse of Σ and $\wp_\Sigma = \Sigma^+\Sigma$ is the Euclidean projection onto Σ, then

$$q = \underline{x}'A\underline{x}+2\underline{b}'\underline{x}+c \sim \chi^2(r,\tfrac{1}{2}\delta)$$

if and only if

$$q = (\underline{x}+\underline{t})'\Sigma_\Gamma^+(\underline{x}+\underline{t}) \text{ a.e.} \tag{3.147}$$

where
$$\Gamma = \wp_\Sigma\big[\text{Col}(A)\big], \quad \text{Dim}(\Gamma) = r, \tag{3.148}$$

$$\underline{t} = \Sigma\Sigma_\Gamma^+\big[\Sigma(A\underline{\mu}+\underline{b}) - \underline{\mu}\big] \in \Sigma(\Gamma), \tag{3.149}$$

and
$$\delta = (A\underline{\mu}+\underline{b})'\Sigma\Sigma_\Gamma^+\Sigma(A\underline{\mu}+\underline{b}). \tag{3.150}$$

Proof: If $\underline{\mu} = \underline{\mu}_0+\underline{\mu}_1$, $\underline{\mu}_0 \in \text{Null}(\Sigma)$ and $\underline{\mu}_1 \in \text{Col}(\Sigma)$, then from Lemma 3.16, $B = \Sigma^+\Sigma A\Sigma^+\Sigma = \Sigma_\Gamma^+$ and $\underline{m} = \Sigma^+\Sigma(\underline{b}+A\underline{\mu}_0) \in \Gamma$. Thus

$$\begin{aligned}\underline{m} &= \Sigma_\Gamma^+\Sigma\underline{m} = \Sigma_\Gamma^+\Sigma\Sigma^+\Sigma(\underline{b}+A\underline{\mu}_0)\\ &= \Sigma_\Gamma^+\Sigma(\underline{b}+A\underline{\mu} - A\underline{\mu}_1)\\ &= \Sigma_\Gamma^+\Sigma(\underline{b}+A\underline{\mu} - \Sigma_\Gamma^+\underline{\mu})\end{aligned} \tag{3.151}$$

and
$$\underline{t} = \Sigma\underline{m} = \Sigma\Sigma_\Gamma^+\big[\Sigma(\underline{b}+A\underline{\mu}) - \underline{\mu}\big]. \tag{3.152}$$

In addition
$$\Sigma_\Gamma^+(\underline{\mu}+\underline{t}) = \Sigma_\Gamma^+\Sigma(\underline{\mu}+\underline{b}). \tag{3.153}$$

●

3.9 {2}-Inverses and the Distribution and Independence of Second Degree Polynomials

Thus far in the chapter, the discussion of the distribution of quadratic forms and second degree polynomials has been limited to a characterization of the canonical forms leading to a χ^2 distribution. The unique expression of a second degree polynomial in terms of its First Canonical Form does not provide much insight into their general distribution. In this section a more refined canonical representation in terms of {2}-inverses of Σ is proposed for arbitrary second degree polynomials. In terms of the Second Canonical Form, the distribution of an arbitrary second degree polynomial becomes apparent. As noted in the Introduction, the distribution of an arbitrary second degree polynomial was characterized by Baldessari (1967) assuming nonsingular covariance and by Khatri (1977) assuming nonsingular covariance. The development of a Second Canonical Form in this section borrows from that of Baldessari and Khatri; however, expression of the result in terms of {2}-inverses leads to a simplification not discussed by these authors.

Theorem 3.18: For every second degree polynomial in $\underline{x} \sim \mathcal{N}(\underline{\mu},\Sigma)$

$$q = \underline{x}'A\underline{x} + 2\underline{b}'\underline{x} + c$$

there exists unique

1. nonzero scalars $\lambda_1 > \cdots > \lambda_k$, which are the nonzero eigenvalues of $A\Sigma$ corresponding to the eigenspaces $\mathcal{F}_1, \cdots, \mathcal{F}_k$; (3.154)
2. Σ-orthogonal subspaces $\Gamma_0, \Gamma_1, \cdots, \Gamma_k$ such that $Col(\Sigma) = \Gamma_0 \oplus \Gamma_1 \oplus \cdots \oplus \Gamma_k$, where $\Gamma_i = \Sigma^+\Sigma(\mathcal{F}_i)$ for $i = 1, \cdots, r$; (3.155)
3. vectors $\underline{t}_1, \cdots, \underline{t}_k$ given by

$$\underline{t}_i = \Sigma\Sigma^+_{\Gamma_i}\left[\lambda_i^{-1}\Sigma(A\underline{\mu} + \underline{b}) - \underline{\mu}\right] \varepsilon \Sigma(\Gamma_i); \tag{3.156}$$

4. scalars $\delta_1, \cdots, \delta_k$ given by

$$\delta_i = \lambda_i^{-2}[A\underline{\mu} + \underline{b}]'\Sigma\Sigma^+_{\Gamma_i}\Sigma[A\underline{\mu} + \underline{b}] \tag{3.157}$$

5. a vector given by

$$\underline{b}_0 = \Sigma^+_{\Gamma_0}\Sigma\underline{b} \;\varepsilon\; \Gamma_0 \tag{3.158}$$

such that

$$q = \sum_{i=1}^{k}\left(\lambda_i[\underline{x} + \underline{t}_i]'\Sigma^+_{\Gamma_i}[\underline{x} + \underline{t}_i]\right) + \underline{b}_0'[\underline{x} - \underline{\mu}] + \underline{\mu}'A\underline{\mu} + 2\underline{b}'\underline{\mu} + c - \sum_{i=1}^{k}\lambda_i\delta_i \tag{3.159}$$

where the quadratic forms

$$q^*_{\Gamma_i} = [\underline{x} + \underline{t}_i]'\Sigma^+_{\Gamma_i}[\underline{x} + \underline{t}_i] \tag{3.160}$$

are independently distributed $\chi^2(r_i, \frac{1}{2}\delta_i)$ with $r_i = Dim(\Gamma_i)$.

Proof: For $B = \Sigma^+\Sigma A\Sigma\Sigma^+$ with Σ^+ the Moore-Penrose Inverse of Σ, suppose,

$$A\Sigma\underline{v} = \lambda\underline{v} \tag{3.161}$$

then

$$B\Sigma(\Sigma^+\Sigma\underline{v}) = \Sigma^+\Sigma A\Sigma\underline{v} = \lambda\Sigma^+\Sigma A\underline{v}. \tag{3.162}$$

Since a priori $Rank(B\Sigma) \leq Rank(A\Sigma)$, (3.162) implies the eigenvalues of $A\Sigma$ and $B\Sigma$ are identical with the same multiplicity. If $\mathcal{F}_i$ is the eigenspace of

$A\Sigma$ corresponding to λ_i, then $\Gamma_i = \Sigma^+\Sigma(\mathcal{F}_i)$ is the corresponding eigenspace of $B\Sigma$. In accordance with Corollary 2.9.1, B may be expressed as

$$B = \sum_{i=1}^{k} \lambda_i \Sigma^+_{\Gamma_i}, \tag{3.163}$$

for the Σ-orthogonal spaces $\Gamma_1, \cdots, \Gamma_k$. Let Γ_0 be such that $\mathrm{Col}(\Sigma) = \Gamma_0 \oplus \Gamma_1 \oplus \cdots \oplus \Gamma_k$, where Γ_i and Γ_j Σ-orthogonal, then in addition to (3.163)

$$\Sigma^+ = \sum_{i=0}^{k} \Sigma^+_{\Gamma_i}. \tag{3.164}$$

As from (3.129) in the proof of Lemma 3.12, the following holds almost everywhere

$$q = [\underline{x} - \underline{\mu}]'B[\underline{x} - \underline{\mu}] + 2[A\underline{\mu} + \underline{b}]'\Sigma\Sigma^+[\underline{x} - \underline{\mu}] + \underline{\mu}'A\underline{\mu} + 2\underline{b}'\underline{\mu} + c. \tag{3.165}$$

Therefore, from (3.160) and (3.161),

$$\begin{aligned} q = \sum_{i=1}^{k} \lambda_i \Big\{ & [\underline{x} - \underline{\mu}]'\Sigma^+_{\Gamma_i}[\underline{x} - \underline{\mu}] + 2\lambda_i^{-1}[A\underline{\mu} + \underline{b}]'\Sigma\Sigma^+_{\Gamma_i}[\underline{x} - \underline{\mu}] \\ & + \lambda_i^{-2}[A\underline{\mu} + \underline{b}]'\Sigma\Sigma^+_{\Gamma_i}\Sigma[A\underline{\mu} + \underline{b}]\Big\} \\ & + 2\Big\{[A\underline{\mu} + \underline{b}]'\Sigma\Sigma^+_{\Gamma_0}[\underline{x} - \underline{\mu}]\Big\} + \underline{\mu}'A\underline{\mu} + 2\underline{b}'\underline{\mu} + c \\ & - \sum_{i=1}^{k} \lambda_i^{-1}\Big\{[A\underline{\mu} + \underline{b}]'\Sigma\Sigma^+_{\Gamma_i}[A\underline{\mu} + \underline{b}]\Big\} \end{aligned} \tag{3.166}$$

$$\begin{aligned} = \sum_{i=1}^{k} \lambda_i \Big\{ & \big[\underline{x} - \underline{\mu} + \lambda_i^{-1}\Sigma(A\underline{\mu} + \underline{b})\big]'\Sigma^+_{\Gamma_i}\big[\underline{x} - \underline{\mu} + \lambda_i^{-1}\Sigma(A\underline{\mu} + \underline{b})\big]\Big\} \\ & + 2[A\underline{\mu} + \underline{b}]'\Sigma\Sigma^+_{\Gamma_0}[\underline{x} - \underline{\mu}] + \underline{\mu}'A\underline{\mu} + 2\underline{b}'\underline{\mu} + c \\ & + \sum_{i=0}^{k} \lambda_i^{-1}\Big\{[A\underline{\mu} + \underline{b}]'\Sigma\Sigma^+_{\Gamma_i}\Sigma[A\underline{\mu} + \underline{b}]\Big\}. \end{aligned} \tag{3.167}$$

(3.158) follows since, by definition of Γ_0,

$$\Sigma^+_{\Gamma_0}\Sigma\underline{b} = \Sigma^+_{\Gamma_0}\Sigma[A\underline{\mu} + \underline{b}]. \tag{3.168}$$

From Corollary 3.11.1, the quadratic forms

$$q^*_{\Gamma_i} = (\underline{x} - \underline{t}_i)'\Sigma^+_{\Gamma_i}(\underline{x} + \underline{t}_i) \tag{3.169}$$

are independently distributed χ^2distributed. They are also independent of $\underline{b}_0'\underline{x}$.

•

The quadratic forms $q^*_{\Gamma_i}$ are in a sense the elementary building blocks for the construction of all second degree polynomials. This motivates the following definition.

Definition 3.3: For $\underline{x} \sim \mathcal{N}(\underline{\mu},\Sigma)$, a subspace $\Gamma \subset \mathrm{Col}(\Sigma)$ and $\underline{t} \in \Sigma(\Gamma)$, the χ^2 distributed quadratic form

$$q^*_\Gamma = (\underline{x}+\underline{t})'\Sigma^+_\Gamma(\underline{x}+\underline{t}) \tag{3.170}$$

is an Elementary Quadratic Form of degree $r = \mathrm{Dim}(\Gamma)$.

•

Theorem 3.18 may be restated in terms of elementary quadratic forms as follows:

Corollary 3.18.1: If $\underline{x} \sim \mathcal{N}(\underline{\mu},\Sigma)$ then $q = \underline{x}'A\underline{x}+2\underline{b}'\underline{x}+c$ can be uniquely expressed as

$$q = \sum_{i=1}^{k} \lambda_i q^*_{\Gamma_i} + 2\underline{b}_0'(\underline{x} - \underline{\mu}) + d \ , \tag{3.171}$$

with scalars $\lambda_1 > \cdots > \lambda_k$ and d, and mutually independent elementary quadratic forms $q^*_{\Gamma_1},\cdots,q^*_{\Gamma_k}$ which are each independent of $\underline{b}_0'(\underline{x} - \underline{\mu})$.

•

Definition 3.4: Expression (3.171) in Corollary 3.18.1 is the Second Canonical Form of the second degree polynomial q.

•

As mentioned earlier, the characterization of the distribution of $q = \underline{x}'A\underline{x} + 2\underline{b}'\underline{x}+c$ contained in Theorem 3.16 has appeared in the literature. While the general results are identical, the result in terms of {2}-inverses is computationally more tractable than the approaches previously published. For example, Khatri (1980) expresses the noncentrality parameter as

$$\delta_i = \lambda_i^{-2}[\underline{b}+A\underline{\mu}]'TE_iT'[\underline{b}+A\underline{\mu}], \tag{3.172}$$

where $\Sigma = TT'$ is the full rank factorization and

$$T'AT = \sum_{i=1}^{k} \lambda_i E_i \tag{3.173}$$

is the spectral decomposition. To compute (3.172) from A and Σ, Khatri offers the following expression:

$$TE_jT' = \frac{\left[\Sigma \prod_{\substack{j=0 \\ j\neq i}}^{k} (A\Sigma - \lambda_i)\right]}{\left[\prod_{\substack{j=0 \\ j\neq i}}^{k} (\lambda_i - \lambda_j)\right]} . \tag{3.174}$$

In contrast, (3.157) is far simpler. Not only does the Second Canonical Form represent an improvement over previous respresentations of second degree polynomials, but also, in its terms, independent second degree polynomials may be easily identified.

Theorem 3.19: Let the second degree polynomials

$$q = \underline{x}'A\underline{x} + 2\underline{b}'\underline{x} + c \tag{3.175}$$

and

$$p = \underline{x}'Q\underline{x} + 2\underline{m}'\underline{x} + d \tag{3.176}$$

have their Second Canonical Forms,

$$q = \sum_{i=1}^{k} \lambda_i q^*_{\Gamma_i} + 2\underline{b}_0'(\underline{x} - \underline{\mu}) + c_0 \tag{3.177}$$

$$p = \sum_{i=1}^{n} \upsilon_i q^*_{\Psi_i} + 2\underline{m}_0'(\underline{x} - \underline{\mu}) + d_0 . \tag{3.178}$$

Then p and q are independent if and only if the spaces $\{\underline{b}_0\}$, $\Gamma_1, \cdots, \Gamma_k$, $\{\underline{m}_0\}$, $\Psi_1, \cdots, \Psi_n$, are Σ-orthogonal.

Proof: As in the proof of Lemma 3.12, rewrite p and q almost everywhere as

$$q = [\underline{x} - \underline{\mu}]'B[\underline{x} - \underline{\mu}] + 2[A\underline{\mu} + \underline{b}]'\Sigma\Sigma^+[\underline{x} - \underline{\mu}] + c_1 \tag{3.179}$$

$$p = [\underline{x} - \underline{\mu}]'R[\underline{x} - \underline{\mu}] + 2[Q\underline{\mu} + \underline{b}]'\Sigma\Sigma^+[\underline{x} - \underline{\mu}] + d_1 \tag{3.180}$$

where

$$B = \sum_{i=1}^{k} \lambda_i \Sigma^+_{\Gamma_i}, \tag{3.181}$$

$$R = \sum_{i=1}^{n} \upsilon_i \Sigma^+_{\Psi_i}, \tag{3.182}$$

and
$$\Sigma^{+} = \sum_{i=0}^{k} \Sigma^{+}_{\Gamma_i} = \sum_{i=0}^{n} \Sigma^{+}_{\Psi_i}. \tag{3.183}$$

By Theorem 3.3
$$0 = \Sigma B \Sigma R \Sigma, \tag{3.184}$$
$$0 = \Sigma R \Sigma \Sigma^{+} \Sigma [A\underline{\mu} + \underline{b}], \tag{3.185}$$
$$0 = \Sigma B \Sigma \Sigma^{+} \Sigma [Q\underline{\mu} + \underline{m}], \tag{3.186}$$
$$0 = [A\underline{\mu} + \underline{b}]' \Sigma \Sigma^{+} \Sigma \Sigma^{+} \Sigma [Q\underline{\mu} + \underline{m}], \tag{3.187}$$
Now (3.184) holds if and only if
$$0 = B \Sigma R. \tag{3.188}$$
This in turn implies
$$0 = \Sigma R \Sigma \Sigma^{+} \Sigma [A\underline{\mu} + \underline{b}] = \Sigma R \Sigma \Sigma^{+}_{\Gamma_0} \Sigma \underline{b} = \Sigma R \Sigma \underline{b}_0, \tag{3.189}$$
$$0 = \Sigma B \Sigma \Sigma^{+} \Sigma [Q\underline{\mu} + \underline{m}] = \Sigma R \Sigma \Sigma^{+}_{\Psi_0} \Sigma \underline{m} = \Sigma R \Sigma \underline{m}_0, \tag{3.190}$$
or equivalently
$$0 = R \Sigma \underline{b}_0 = B \Sigma \underline{m}_0. \tag{3.191}$$
Together (3.188) and (3.189) imply the Σ-orthogonality of the spaces $\{\underline{b}_0\}$, $\{\underline{m}_0\}$, $\Gamma_1, \cdots, \Gamma_k$, $\Psi_1, \cdots, \Psi_n$.

●

CHAPTER IV
{2}-INVERSES AND LEAST SQUARE SOLUTIONS

4.0 Introduction

The concept of least squares is the cornerstone of regression and linear model analysis. Many statistical texts consider at length the associated theory and propose various strategies for obtaining least squares solutions (LSS's), e.g. Searle (1971), Graybill (1976), and Draper and Smith (1981).

The usual approach for obtaining LSS's leads to the normal equations which can be solved by an application of a g-inverse. LSS's may not be unique but can be made so by requiring the solution to the normal equations to satisfy an additional set of nonestimable constraints of appropriate rank. The relationship between these two solution methods has been explored in papers by Mazumdar et al. (1980) and by Searle (1984), who each propose an algorithm for finding a g-inverse yielding a LSS satisfying a specified set of constraints.

Also of interest are estimable constraints imposed on the LSS either alone, as in hypothesis testing, or together with nonestimable constraints, as in hypothesis testing in constrained linear models. This latter problem was considered by Hackney (1976).

Independent of statistical applications, the problem of obtaining LSS's subject to arbitrary constraints, regardless of estimability and rank, has been explored extensively in the literature by Rao and Mitra (1971, 1973), by Ben-Israel and Greville (1974) and more recently by Golub and Van Loan (1983). Considered also in these references is the problem of identifying the LSS having the smallest norm. In general the solutions proposed by these authors are not easily implemented nor do they provide insight into the nature of the LSS's.

An excellent treatment of constrained and minimum norm LSS's is that of

Rao and Mitra (1973), where they approach the problem of solving for LSS's in terms of constrained inverses. As discussed in Chapter II, the notion of constrained inverses is the genesis of {2}-inverses. Thus the paper of Rao and Mitra (1973) strongly suggests the important role of {2}-inverses in least squares theory. In this chapter the utility of Bott-Duffin Inverses in generating LSS's satisfying arbitrary constraints will be explored. Such an approach will be shown to have merit in that it is computationally simple and geometrically motivated. In the course of the chapter the relationships among LSS's in the regression setting which were outlined by Monlezun and Speed (1980) will be extended to the less than full rank case.

In developing this construction, the notion of a minimal sufficient set of matrices as the minimal set of matrices sufficient to generate all LSS's will be introduced. The class of Bott-Duffin Inverses will be shown to be one such minimal sufficient set of matrices.

The outline of the next ten sections of this chapter is as follows.

4.1 The problem of finding constrained and unconstrained LSS's is posed in a series of definitions which establishes the notation for the remainder of the chapter.

4.2 Two common approaches for obtaining solutions to the normal equations $X'X\underline{\beta} = X'\underline{y}$ are identified: (1) choosing a g-inverse of $X'X$ and (2) imposing a complete set of nonestimable constraints. For approach (1), the class of symmetric {1,2}-inverses of $X'X$ is shown to be the minimal sufficient set of g-inverses for generating all solutions to the normal equations. As a consequence, all LSS's may be thought of as indexed by certain subspaces which are disjoint from the nullity of X. This leads to the notion of the Γ-LSS given in Definition 4.2.

4.3 In Lemma 4.2, the one-to-one correspondence between the set of nonestimable constraints imposed on the normal equations and the set of symmetric {1,2}-inverses of $X'X$ is established.

4.4 Symmetric {1,2}-inverses of $X'X$ are a subset of the Bott-Duffin Inverses. In an extension of the results of the previous section, Theorem 4.3 establishes the class of Bott-Duffin Inverses of $X'X$ as the minimal sufficient set of matrices for generating all constrained LSS's.

4.5 As g-inverses are to the solutions to the normal equations so are (1.5)-inverses to constrained LSS's. This correspondence is detailed in this section.

4.6 The relationships existing between constained LSS's established by Monlezun and Speed (1980) in the regression setting are extended to the less than full rank case. In particular, for a pair of nested constaints, Corollary 4.3.4 establishes a direct sum decomposition of the associated LSS's.

4.7 Lemma 4.4 gives a Bott-Duffin expression for the unique LSS satisfying a specified constraint and also having the smallest N-norm.

4.8 Theorem 4.5 extends the results of Sections 4.4 and 4.7 to the LSS's minimizing a weighted sum of squares and satisfying a set of homogeneous constaints.

4.9 Lemma 4.6 gives, in terms of Bott-Duffin Inverses, an expression for the residual sum of squares associated with a particular LSS. Also given is a simplified expression for the difference between the residual sum of squares associated with a pair of nested constaints.

4.10 The results of Sections 4.4 and 4.9 are applied in proposing a computational procedure for calculating the LSS and the corresponding residual sum of squares for any specified constraint. The procedure, based on the G2SWEEP operator of Goodnight (1979), does not depend upon the factorization of X or solving for its eigenvectors.

4.1 The Least Squares Problem

In this section the problem of finding LSS's in the general setting is posed. Since the results of this chapter will be referenced in the next chapter where their application to linear models is considered, it is convenient to establish a notation which is common to this particular application. However, the results of this chapter are quite general. We will begin with some definitions:

Definition 4.1:

1. $\hat{\beta} = A\underline{y}$ is a Least Squares Solution if $\hat{\beta}$ minimizes the Weighted Sum of Squares

$$\|\underline{y} - X\beta\|_W^2 = [\underline{y} - X\beta]'W[\underline{y} - X\beta], \tag{4.1}$$

where W is a known nxn n.n.d. matrix,

X is a known nxp matrix of rank r,

$\underline{y}$ is an nx1 vector of constants, and

$\underline{\beta}$ is a px1 vector of parameters.

2. $\hat{\underline{\beta}} = A\underline{y}$ is a Constrained Least Squares Solution if $\hat{\underline{\beta}}$ minimizes (4.1) subject to a consistent set of constraints

$$H'\underline{\beta} = \underline{h} \,, \tag{4.2}$$

where H′ is a known cxp matrix of rank c, and

$\underline{h}$ is a cx1 vector of constants.

3. The LSS $\hat{\underline{\beta}}$ is a Minimum N-Norm Least Squares Solution if it has the smallest N-norm of all LSS's,

$$\|\hat{\underline{\beta}}\|_N^2 = \hat{\underline{\beta}}'N\hat{\underline{\beta}} \,, \tag{4.3}$$

where N is a known pxp positive definite matrix.

4. The LSS $\hat{\underline{\beta}}$ is a Minimum N-Norm Constrained Least Squares Solution if it has the smallest N-norm of all constrained LSS's which satisfy a specified set of constraints.

5. The minimized value of (4.1), the Residual Sum of Squares, will be denoted by

$$\min_{\underline{\beta}} \|\underline{y} - X\underline{\beta}\|_W^2 \tag{4.4}$$

or subject to (4.2) by

$$\min_{H'\underline{\beta}=\underline{h}} \|\underline{y} - X\underline{\beta}\|_W^2 \,. \tag{4.5}$$

•

Two types of constraints are of particular interest in linear model analysis and will be specifically considered when developing constrained LSS's:

Definition 4.2:

1. $L'\underline{\beta} = 0$ is a set of estimable constraints if $\text{Row}(L') \subset \text{Row}(X)$.
2. $K'\underline{\beta} = 0$ is a set of nonestimable constraints if $\text{Row}(K') \cap \text{Row}(X) = \{\underline{0}\}$.

•

4.2 Strategies for Obtaining Least Square Solutions

For this, and the next few sections, the discussion will be limited to obtaining constrained LSS's where $W = I$ and $\underline{h} = \underline{0}$. In this case, the LSS's minimizing $\|\underline{y} - X\underline{\beta}\|^2$ may be found as solutions to the normal equations

$$X'X\hat{\underline{\beta}} = X'\underline{y}. \tag{4.6}$$

When X is of full column rank (i.e. p = r) the solution to (4.6) is unique and given by

$$\hat{\underline{\beta}} = (X'X)^{-1}X'\underline{y}. \tag{4.7}$$

When X is less than full column rank there are numerous solutions to the equations. In this case there are two common strategies employed to obtain a solution:

METHOD I: Choose a g-inverse (i.e. a {1}-inverse) of X'X, say G, and then form the LSS $\hat{\underline{\beta}} = GX'\underline{y}$;

METHOD II: Impose a set of p-r linearly independent nonestimable constraints, $K'\underline{\beta} = \underline{0}$, in addition to the normal equations and solve for the unique LSS satisfying both $X'X\underline{\beta} = X'\underline{y}$ and $K'\underline{\beta} = \underline{0}$.

The linkage between the two methods has been explored in the literature. Mazumdar, Li and Bryce (1980), and Searle (1984) constructed g-inverses of X'X such that the associated solution statisfies a specified set of nonestimable constraints. This linkage will now be further explored.

With Method I, it is redundant to consider all possible g-inverses of X'X since different g-inverses, G_1 and G_2, may yield identical LSS's for all possible values of $\underline{y}$. To remove this redundancy we need to characterize the equivalence classes defined by

$$G_1 \equiv G_2 \quad \text{if and only if} \quad \hat{\underline{\beta}}_1 = G_1X'\underline{y} = G_2X'\underline{y} = \hat{\underline{\beta}}_2, \text{ for all } \underline{y}. \tag{4.8}$$

It is not readily apparent when two g-inverses G_1 and G_2 belong to the same equivalence class, since, for a particular choice of $\underline{y}$, it may be true that $G_1X'\underline{y} = G_2X'\underline{y}$, but this equality may not hold for all values of $\underline{y}$. This is illustrated in the following example from linear models.

Example 4.1: Consider the balanced one-way ANOVA

DATA

a_1	a_2	a_+
b_1	b_2	b_+

(4.9)

and the corresponding normal equations

$$\begin{bmatrix} 4 & 2 & 2 \\ 2 & 2 & 0 \\ 2 & 0 & 2 \end{bmatrix} \begin{bmatrix} \mu \\ \alpha_1 \\ \alpha_2 \end{bmatrix} = \begin{bmatrix} a_+ + b_+ \\ a_+ \\ b_+ \end{bmatrix}. \tag{4.10}$$

It is trivial to verify that the following are g-inverses of X′X.

$$G_1 = \frac{1}{2} \begin{bmatrix} 0 & 0 & 0 \\ 0 & 1 & 0 \\ 0 & 0 & 1 \end{bmatrix}, \tag{4.11}$$

$$G_2 = \frac{1}{4} \begin{bmatrix} 1 & -1 & -1 \\ 0 & 2 & 0 \\ 0 & 0 & 2 \end{bmatrix}, \tag{4.12}$$

and

$$G_3 = \frac{1}{2} \begin{bmatrix} 5 & -6 & -3 \\ -6 & 8 & 4 \\ -3 & 4 & 2 \end{bmatrix}. \tag{4.13}$$

Let $\hat{\underline{\beta}}_i = G_i X' \underline{y}$ for i = 1, 2, 3. For the data

DATA

14	6	20
6	4	10

(4.14)

$$\hat{\underline{\beta}}_i = \begin{bmatrix} 0 \\ 10 \\ 5 \end{bmatrix} \text{ for } i = 1, 2, 3. \tag{4.15}$$

However, for the data

DATA

4	2	6
5	3	8

(4.16)

$$\hat{\underline{\beta}}_i = \begin{bmatrix} 0 \\ 3 \\ 4 \end{bmatrix} \text{for } i = 1, 2 \tag{4.17}$$

and

$$\hat{\underline{\beta}}_3 = \begin{bmatrix} 5 \\ -2 \\ -1 \end{bmatrix}. \tag{4.18}$$

Given any a_1, a_2, b_1, b_2 in (4.9) it can be verified that G_1 and G_2 will always yield the identical LSS's satisfying $\mu = 0$ while G_3 will always lead to a LSS satisfying $a_2 - 2a_3 = 0$.

●

The above example shows that it is possible to identify equivalent g-inverses. The following lemma establishes the criteria for determining when two g-inverses are equivalent. As a corollary to the lemma, a single representative of each class is identified.

Lemma 4.1: Consider the g-inverses

$$G_1 = (X'X)^+_{\mathcal{S},\mathcal{F}} + Z \tag{4.19}$$

and

$$G_2 = (X'X)^+_{\Gamma,\Psi} + Z_* \; . \tag{4.20}$$

Then

$$G_1X'\underline{y} = G_2X'\underline{y}, \text{ for all } \underline{y} \tag{4.21}$$

if and only if

$$(X'X)^+_{\mathcal{S}} = G_1X'XG_1' = G_2X'XG_2' = (X'X)^+_{\Gamma} \; . \tag{4.22}$$

Proof: From Corollary 2.3.1

$$G_1X' = (X'X)^+_{\mathcal{S},\mathcal{F}}X' \tag{4.23}$$

and

$$G_2X' = (X'X)^+_{\Gamma,\Psi}X'. \tag{4.24}$$

(⇒) (4.23) and (4.24) together imply that for all $\underline{z}$

$$(X'X)^+_{\mathcal{S},\mathcal{F}}X'X\underline{z} = (X'X)^+_{\Gamma,\Psi}X'X\underline{z} \; . \tag{4.25}$$

Thus $\mathcal{S} = \Gamma$, or equivalently (4.22).

(⇐) Since (4.22) is equivalent to $\mathcal{S} = \Gamma$, the result follows from (4.23) and (4.24).

●

Corollary 4.1.1: For any g-inverse of X′X

$$G = (X'X)^{+}_{\Gamma,\Psi} + Z \tag{4.26}$$

the following equality holds for all $\underline{y}$:

$$GX'\underline{y} = (X'X)^{+}_{\Gamma}X'\underline{y}\ . \tag{4.27}$$

●

As a consequence of Corollary 4.1.1, the set of all symmetric {1,2}-inverses is sufficient to generate all solutions to the normal equations. Furthermore, the set is minimal in that no two symmetric {1,2}-inverses generate identical LSS's for all values of $\underline{y}$. Therefore, the set of all symmetric {1,2}-inverses of X′X is a minimal sufficient set of g-inverses generating all solutions to the normal equations. In Section 4.4, it will be shown that the set of Bott-Duffin Inverses is the minimal sufficient set of generalized inverses for generating all constrained LSS's. It will be convenient to identify the LSS generated by a particular Bott-Duffin Inverse. This necessitates the following definition:

Definition 4.2: For a Bott-Duffin Inverse $(X'X)^{+}_{\Gamma}$ of X′X,

$$\hat{\underline{\beta}}_{\Gamma} = (X'X)^{+}_{\Gamma}X'\underline{y} \tag{4.28}$$

is a Γ-Least Squares Solution (Γ-LSS).

●

To avoid any misconception, it should be understood that even though $\Gamma \neq \Psi$, for a particular $\underline{y}$ it may happen that $\hat{\underline{\beta}}_{\Gamma} = \hat{\underline{\beta}}_{\Psi}$. For example, if $\underline{z} \in \Gamma \cap \Psi$ and $\underline{y} = X\underline{z}$ then

$$\hat{\underline{\beta}}_{\Gamma} = \underline{z} = \hat{\underline{\beta}}_{\Psi}. \tag{4.29}$$

Nevertheless, $\hat{\underline{\beta}}_{\Gamma}$ and $\hat{\underline{\beta}}_{\Psi}$ considered as functions of $\underline{y}$ will differ at some point over the values of $\underline{y}$ whenever $\Gamma \neq \Psi$.

4.3 Symmetric {1,2}-Inverses And Sets Of Nonestimable Constraints

Example 4.1 suggests that each equivalence class, and hence each symmetric {1,2}-inverse, is associated with a unique set of p-r linearly independent nonestimable constraints. This one-to-one correspondence is formally established in the following lemma.

Lemma 4.2: $\hat{\underline{\beta}}$ is the unique LSS satisfying the set of p - r linearly independent nonestimable constraints $K'\hat{\underline{\beta}} = 0$ if and only if $\hat{\underline{\beta}} = \hat{\underline{\beta}}_{\Gamma}$ where

Γ is the orthogonal complement of Col(K).

Proof:

($\Rightarrow$) Write K as $K = K_0 + K_1$ where $Col(K_0) = Null(X)$ and $Col(K_1) \subset Row(X)$. If $\underline{\gamma} \varepsilon \Gamma \cap Null(X)$ then

$$\underline{0} = K'\underline{\gamma} = K_0'\underline{\gamma} + K_1'\underline{\gamma} = K_0'\underline{\gamma} \tag{4.30}$$

which is a contradicton unless $\underline{\gamma} = \underline{0}$. Thus $\hat{\underline{\beta}}_\Gamma$ is well defined and $K'\hat{\underline{\beta}}_\Gamma = \underline{0}$.

($\Leftarrow$) For an r-dimensional subspace Γ such that $\Gamma \cap Null(X) = \{\underline{0}\}$, let G be a full column rank matrix such that $\Gamma = Col(G)$. Write G as $G = G_0 + G_1$ where $Col(G_0) \subset Null(X)$ and $Col(G_1) = Row(X)$. Suppose $\underline{\upsilon} \varepsilon Row(X)$ is orthogonal to Γ, then

$$\underline{0} = G'\underline{\upsilon} = G_0'\underline{\upsilon} + G_1'\underline{\upsilon} = G_1'\underline{\upsilon}\ , \tag{4.31}$$

which is a contradiction unless $\underline{\upsilon} = \underline{0}$. Thus, $K'\underline{\beta} = \underline{0}$ is a non-estimable set of constraints and $K'\hat{\underline{\beta}}_\Gamma = \underline{0}$.

●

Nonestimable constraints are often used to obtain a solution vector satisfying particular constraints. As such, their use is usually viewed as a matter of convenience, but not as a matter of necessity. On the other hand, the use of g-inverses is viewed as necessary and sufficient to obtain a solution. The above lemma indicates that, whether one realizes it or not, by choosing a g-inverse to obtain a solution, one is in fact imposing a set of constraints on the solution. Since the orthogonal complement of a subspace can be obtained either by the Gram-Schmidt orthogonalization process or by inspection, Lemma 4.2 offers a straightforward construction of a symmetric {1,2}-inverse of X'X yielding a LSS satisfying a given set of constraints. We illustrate this in the following example, again from linear models:

Example 4.2: Consider fitting

$$y_{ij} = m + a_i + b_j + e_{ij} \tag{4.32}$$

to the data

DATA

18 15	16	49
23	12	35
56	28	84

. (4.33)

The normal equations are

$$\begin{bmatrix} 5 & 3 & 2 & 3 & 2 \\ 3 & 3 & 0 & 2 & 1 \\ 2 & 0 & 2 & 1 & 1 \\ 3 & 2 & 1 & 3 & 0 \\ 2 & 1 & 1 & 0 & 2 \end{bmatrix} \begin{bmatrix} \mu \\ \alpha_1 \\ \alpha_2 \\ \beta_1 \\ \beta_2 \end{bmatrix} = \begin{bmatrix} 84 \\ 49 \\ 35 \\ 56 \\ 28 \end{bmatrix}. \tag{4.34}$$

To obtain a solution corresponding to $a_1 + a_2 = 0$ and $b_1 + b_2 = 0$, let

$$K_1' = \begin{bmatrix} 0 & 1 & 1 & 0 & 0 \\ 0 & 0 & 0 & 1 & 1 \end{bmatrix}. \tag{4.35}$$

The vector space orthogonal to Row(K_1') has as a basis matrix

$$T_1 = \begin{bmatrix} 1 & 0 & 0 \\ 0 & 1 & 0 \\ 0 & -1 & 0 \\ 0 & 0 & 1 \\ 0 & 0 & -1 \end{bmatrix}. \tag{4.36}$$

Using this we obtain

$$(X'X)^+_{\Gamma_1} = \frac{1}{28} \begin{bmatrix} 6 & -1 & 1 & -1 & 1 \\ -1 & 6 & -6 & -1 & 1 \\ 1 & -6 & 6 & 1 & -1 \\ -1 & -1 & 1 & 6 & 6 \\ 1 & 1 & -1 & -6 & 6 \end{bmatrix}. \tag{4.37}$$

This leads to the solution

$$\hat{\underline{\beta}}_{\Gamma_1} = (X'X)^+_{\Gamma_1} X'\underline{y} = \frac{1}{2} \begin{bmatrix} 33 \\ -2 \\ 2 \\ 5 \\ -5 \end{bmatrix}. \tag{4.38}$$

Similarly, to obtain a solution corrsponding to $\alpha_2 = 0$ and $\beta_2 = 0$, let

$$K_2' = \begin{bmatrix} 0 & 0 & 1 & 0 & 0 \\ 0 & 0 & 0 & 0 & 1 \end{bmatrix}. \tag{4.39}$$

The vector space orthogonal to Row(K_2') has as a basis matrix

$$T_2 = \begin{bmatrix} 1 & 0 & 0 \\ 0 & 1 & 0 \\ 0 & 0 & 0 \\ 0 & 0 & 1 \\ 0 & 0 & 0 \end{bmatrix}. \tag{4.40}$$

Using this we obtain

$$(X'X)^{+}_{\Gamma_2} = \frac{1}{7}\begin{bmatrix} 5 & -3 & 0 & -3 & 0 \\ -3 & 6 & 0 & -1 & 0 \\ 0 & 0 & 0 & 0 & 0 \\ -3 & -1 & 0 & 6 & 0 \\ 0 & 0 & 0 & 0 & 0 \end{bmatrix} \tag{4.41}$$

and

$$\hat{\underline{\beta}}_{\Gamma_2} = (X'X)^{+}_{\Gamma_2}X'\underline{y} = \begin{bmatrix} 15 \\ -2 \\ 0 \\ 5 \\ 0 \end{bmatrix}. \tag{4.42}$$

•

4.4 Bott-Duffin Inverses And Constrained LSS's

When the design matrix X is full column rank, the LSS minimizing $\| \underline{y} - X\underline{\beta}\|^2$ subject to the estimable constraints $L'\underline{\beta} = \underline{0}$ is uniquely given by

$$\hat{\underline{\beta}} = \left\{(X'X)^{-1} - (X'X)^{-1}\left\{L\left[L'(X'X)^{-1}L\right]^{-1}L'\right\}(X'X)^{-1}\right\}X'\underline{y} \tag{4.43}$$

where L is full column rank, Searle [(1971), p. 191]. This expression provides little insight into the nature of the solution and computationally is intractable. However, an elegant expression in terms of {2}-inverses is given by Hsuan et al. (1985). The derivation is quite simple and requires writing (4.43) as

$$\hat{\underline{\beta}} = \left\{(X'X)^{-1} - (X'X)^{-1}\left[(X'X)^{-1}\right]^{+}_{\Lambda}(X'X)^{-1}\right\}X'\underline{y} \tag{4.44}$$

where $\Lambda = \text{Col}(L)$. Since from (2.175) in Chapter II

$$\left(A^{-1}\right)^{+}_{\mathcal{S}^{\perp}} = A - AA^{+}_{\mathcal{S}}A, \tag{4.45}$$

(4.44) may be rewritten as

$$\hat{\underline{\beta}}_{\Gamma} = (X'X)^{+}_{\Gamma}X'\underline{y}\ , \tag{4.46}$$

where $\Gamma = \text{Null}(L')$. Since $\text{Dim}(\Gamma) < \text{Rank}(X)$, $(X'X)^{+}_{\Gamma}$ in (4.56) is strictly a {2}-inverse of $X'X$. This is in contrast to Corollary 4.1.1 where the LSS to the normal equations was given in terms of a {1,2}-inverse. Thus with this difference the above expression may be considered an extension of Corollary 4.1.1 to the case of estimable constraints in the regression setting. This result holds in general for all constraints, estimable or otherwise, regardless of Rank(X). The following theorem establishes the nitch held by {2}-inverses in generating LSS's.

<u>Theorem 4.3</u>: Given the space Γ, where $\Gamma \not\subset \text{Null}(X)$, then

$$\hat{\underline{\beta}} = A\underline{y} \text{ is a LSS satisfying } \underline{\beta} \in \Gamma$$
$$\text{if and only if} \tag{4.47}$$
$$\hat{\underline{\beta}} = \hat{\underline{\beta}}_{\Psi} = (X'X)^{+}_{\Psi}X'\underline{y}\ ,$$

where $\Psi \subset \Gamma$, $\wp_{\Omega}(\Psi) = \wp_{\Omega}(\Gamma)$, and $\Psi \cap \text{Null}(X) = \{\underline{0}\}$, for $\wp_{\Omega}$ the Euclidean projection onto $\Omega = \text{Row}(X)$.

<u>Proof</u>:

($\Rightarrow$) The following equality holds:

$$\min_{\underline{\beta}\in\Gamma}\left\|\underline{y} - X\underline{\beta}\right\|^2 = \min_{\underline{y}_+ \in X(\Psi)}\left\|\underline{y} - \underline{y}_+\right\|^2 \tag{4.48}$$

Now the LSS for $\underline{y}_+$ is given by $\hat{\underline{y}}_+ = X(X'X)^{+}_{\Psi}X'\underline{y}$, since $X(X'X)^{+}_{\Psi}X'$ is the Euclidean projection onto $X(\Psi) = X(\Gamma)$. Furthermore, since $\Psi \subset \Gamma$,

$$\hat{\underline{\beta}}_{\Psi} = (X'X)^{+}_{\Psi}\underline{y} \in \Gamma \tag{4.49}$$

is a LSS for $\underline{\beta}$ satisfying the constraint.

($\Leftarrow$) If $\hat{\underline{\beta}} \in \Gamma$ is any constrained LSS for $\underline{\beta}$ then, for all $\underline{y}$

$$XA\underline{y} = X(X'X)^{+}_{\Psi}X'\underline{y}. \tag{4.50}$$

and

$$\left(\hat{\underline{\beta}}_{\Psi} - \hat{\underline{\beta}}\right) \in \text{Null}(X) \cap \Gamma. \tag{4.51}$$

This leads to the two cases.

1. $G \cap \text{Null}(X) = \{\underline{0}\}$, in which case $\hat{\underline{\beta}}_{\Psi} = \hat{\underline{\beta}}$ is unique.
2. $G \cap \text{Null}(X) \neq \{\underline{0}\}$, in which case

$$\hat{\underline{\beta}} = (X'X)^{+}_{\Psi}X'\underline{y} + B\underline{y} \tag{4.52}$$

for B with Col(B) $\subset$ Null(X)$\cap$G. Now if Φ = Col$\left[(X'X)^{+}_{\Psi}X' + B\right]$ then $\hat{\underline{\beta}}$ is a LSS subject to $\underline{\beta} \in \Phi \subset \Gamma$. Since $\Psi \cap$ Null(X) = $\{\underline{0}\}$, it follows that $\Phi \cap$ Null(X) = $\{\underline{0}\}$. Thus $\hat{\underline{\beta}}$ is unique. This establishes that for all $\underline{y}$

$$\hat{\beta} = \hat{\underline{\beta}}_{\Phi} = (X'X)^{+}_{\Phi}X'\underline{y}. \tag{4.53}$$

●

The above theorem indicates that any LSS is given in terms of a suitable Bott-Duffin Inverse of X′X. Furthermore, similar to the proof of Lemma 4.1, no two Bott-Duffin Inverses generate the same constrained LSS for all values of y. Consequently, the class of Bott-Duffin Inverses of X′X is the <u>minimal sufficient</u> set of matrices generating all LSS's, constrained or otherwise.

As a specific application of the above theorem, consider the LSS's corresponding to the estimable constraints $L'\hat{\underline{\beta}} = 0$.

<u>Corollary 4.3.1</u>: If Λ = Col(L) $\subset$ Row(X), then all LSS's satisfying $L'\underline{\beta} = 0$ are given by $\hat{\underline{\beta}}_{\Gamma} = (X'X)^{+}_{\Gamma}X'\underline{y}$ where $\Gamma \perp \Lambda$, Dim(Γ) = p - Dim(Λ) and $\Gamma \cap$ Null(X) = $\{\underline{0}\}$. For any particular such Γ, $\hat{\underline{\beta}}_{\Gamma}$ also satisfies the p-r nonestimable constraints $K'\underline{\beta} = \underline{0}$ where Col(K) is the orthogonal complement of $\Gamma \oplus \Lambda$.

The above corollary shows that the LSS satisfying a set of estimable constraints is not unique but can be made so by imposing an additional set of p-r nonestimable constraints.

4.5 <u>{1.5}-Inverses and LSS's</u>

Although all constrained LSS's can be written in terms of Bott-Duffin Inverses it is entirely possible for a matrix G, not a Bott-Duffin Inverse of X′X, that

$$\hat{\underline{\beta}} = GX'\underline{y} \tag{4.54}$$

can be a constrained LSS for all values of $\underline{y}$. For (4.54) to be a solution to the normal equations, G must be a {1}-inverse of X′X. In what sense is this true, if at all, for constrained LSS's? The following corollary characterizes these matrices as {1.5}-inverses of $X'X$.

Corollary 4.3.2: $\hat{\beta} = GX'y$ is a LSS satisfying $\hat{\beta} \in \Gamma$ if and only if G is the {1.5}-inverse of $X'X$

$$G = (X'X)^{+}_{\Psi,\Phi} + Z \tag{4.55}$$

where $\Psi \subset \Gamma$, $\wp_{\Omega}(\Psi) = \wp_{\Omega}(\Gamma)$, $ZX' = 0$, $Z'X'X(\Phi) = \{\underline{0}\}$ and $\wp_{\Omega}$ the Euclidean projection onto $\Omega = \mathrm{Row}(X)$.

Proof:

($\Rightarrow$) For all $\underline{y}$

$$(X'X)^{+}_{\Psi,\Phi}X'\underline{y} = (X'X)^{+}_{\Psi}X'\underline{y} \tag{4.56}$$

and

$$GX'\underline{y} = (X'X)^{+}_{\Psi}X'\underline{y}. \tag{4.57}$$

($\Leftarrow$) From the theorem, there exists a Ψ satisfying the conditions of the lemma such that for all $\underline{y}$

$$GX'\underline{y} = (X'X)^{+}_{\Psi}X'\underline{y}. \tag{4.58}$$

In particular,

$$GX'X\underline{\psi} = (X'X)^{+}_{\Psi}X'X\underline{\psi} = \underline{\psi}$$
$$\text{if and only if} \tag{4.59}$$
$$\underline{\psi} \in \Psi.$$

Thus G is a generalized inverse of $X'X$ of the form (4.55). Since (4.56) holds for all $\underline{y}$, it follows that for all $\underline{y}$

$$\underline{0} = ZX'\underline{y} \tag{4.60}$$

or equivalently $ZX' = 0$.

●

The class of matrices defined in the above corollary is the class of nonnull augmented generalized inverses where in the reduction phase Ψ is chosen to be a subspace of Γ and in the augmentation phase the linear transformation Z is chosen to map Row(X) onto $\{\underline{0}\}$.

4.6 Relationships Among LSS's

As mentioned in the Introduction, the relationships among the least square solutions were given in terms of nonorthogonal projections by Monlezun and Speed (1980) who focused on the case where the design matrix X is full column rank. In this section, these relationships among LSS's will be extended to the less than full rank case in expressions in terms of Bott-

Duffin Inverses. This approach is computationally more tractable than the expression solely in terms of orthogonal projectors.

Corollary 4.3.3: Let Ω = Row(X) and $\wp_{\Omega}$ be the Euclidean projection to Ω. If Γ_1 and Γ_2 are subspaces disjoint from Null(X) such that $\wp_{\Omega}(\Gamma_1) \subset \wp_{\Omega}(\Gamma_2)$ then

$$\hat{\beta}_{\Gamma_1} = (X'X)^{+}_{\Gamma_1} X'X\hat{\beta}_{\Gamma_2} = \wp_{\Gamma_1|\mathcal{N}}\hat{\beta}_{\Gamma_2}, \tag{4.61}$$

where $\mathcal{N} = \{\underline{\eta} \mid \underline{\gamma}'X'X\underline{\eta} = 0 \text{ for all } \underline{\gamma} \in \Gamma_1\}$.

Proof: It can be easily verified that

$$(X'X)^{+}_{\Gamma_1} X'X(X'X)^{+}_{\Gamma_2} X' = (X'X)^{+}_{\Gamma_1} X'. \tag{4.62}$$

The remainder of the corollary follows from Theorem 2.6.

•

A specific consequence of the above corollary is the case of LSS's of the normal equations. In this case, when Dim(Γ) = Rank(X) then $\mathcal{N}$ = Null(X). Thus, $(X'X)^{+}_{\Gamma}X'X = \wp_{\Gamma|\mathcal{N}}$ will be the same for all design matrices whose row spaces are identical. In particular, $(X'X)^{+}_{\Gamma}X'X = (X_*' X_*)^{+}_{\Gamma}X_*' X_*$ where X_* is the matrix formed from X by deleting duplicate rows. An application of this observation will be given in the next chapter.

In addition to the relationship among LSS's contained in the above corollary, other interesting relationships exist among the LSS's which are an immediate consequence of Theorem 4.3. In particular, it is often of interest to obtain a series of LSS's satisfying an increasing set of nested constraints, e.g. as those associated with the reduction in sum of squares. The following corollary gives such a relationship.

Corollary 4.3.4: Let $\Gamma_1 \subset \Gamma_2$ be subspaces disjoint from Null(X) then

$$\hat{\beta}_{\Gamma_2} = \hat{\beta}_{\Gamma_1} + \hat{\beta}_{\Gamma_{2*1}} \tag{4.63}$$

where $\Gamma_{2*1} = \{\underline{\gamma}_2 \in \Gamma_2 \mid \underline{\gamma}_2'X'X\underline{\gamma}_1 = 0 \text{ for all } \underline{\gamma}_1 \in \Gamma_1\}$.

Proof: By Corollary 2.8.3

$$(X'X)^{+}_{\Gamma_2} = (X'X)^{+}_{\Gamma_1} + (X'X)^{+}_{\Gamma_{2*1}}. \tag{4.64}$$

The remainder of the corollary follows immediately.

•

The following example, a continuation of Example 4.2, illustrates

Corollaries 4.3.3 and 4.3.4.

Example 4.2 (continued): To obtain a solution corresponding to the estimable constraint $\beta_1 - \beta_2 = 0$, let

$$K' = \begin{bmatrix} 0 & 0 & 0 & 1 & -1 \end{bmatrix}. \tag{4.65}$$

The space $\Gamma_3 = \text{Col}(T_3)$ where

$$T_3 = \begin{bmatrix} 1 & 0 \\ 0 & 1 \\ 0 & -1 \\ 0 & 0 \\ 0 & 0 \end{bmatrix} \tag{4.66}$$

is orthogonal to Col(K) and disjoint from Null(X). For this space

$$(X'X)^+_{\Gamma_3} = \frac{1}{24}\begin{bmatrix} 5 & -1 & 1 & 0 & 0 \\ -1 & 5 & -5 & 0 & 0 \\ 1 & -5 & 5 & 0 & 0 \\ 0 & 0 & 0 & 0 & 0 \\ 0 & 0 & 0 & 0 & 0 \end{bmatrix}, \tag{4.67}$$

$$\wp_{\Gamma_3|\mathcal{N}} = (X'X)^+_{\Gamma_3}(X'X) = \frac{1}{12}\begin{bmatrix} 12 & 6 & 6 & 7 & 5 \\ 0 & 6 & -6 & 1 & -1 \\ 0 & -6 & 6 & -1 & 1 \\ 0 & 0 & 0 & 0 & 0 \\ 0 & 0 & 0 & 0 & 0 \end{bmatrix} \tag{4.68}$$

and

$$\hat{\beta}_{\Gamma_3} = (X'X)^+_{\Gamma_3}X'\underline{y} = \frac{1}{12}\begin{bmatrix} 203 \\ -7 \\ 7 \\ 0 \\ 0 \end{bmatrix}. \tag{4.69}$$

It is easy to verify that for $\hat{\beta}_{\Gamma_1}$ and $\hat{\beta}_{\Gamma_2}$ given by (4.38) and (4.42) respectively

$$\hat{\beta}_{\Gamma_3} = \wp_{\Gamma_3|\mathcal{N}}\hat{\beta}_{\Gamma_1} = \wp_{\Gamma_3|\mathcal{N}}\hat{\beta}_{\Gamma_2}. \tag{4.70}$$

If $\Gamma_4 = \mathrm{Col}(T_4)$, where

$$T_4 = \begin{bmatrix} 0 \\ 0 \\ 1 \\ 2 \\ -4 \end{bmatrix}, \tag{4.71}$$

then

$$(X'X)^+_{\Gamma_4} = \frac{1}{42}\begin{bmatrix} 0 & 0 & 0 & 0 & 0 \\ 0 & 0 & 0 & 0 & 0 \\ 0 & 0 & 1 & 2 & -4 \\ 0 & 0 & 2 & 4 & -8 \\ 0 & 0 & -4 & -8 & 16 \end{bmatrix} \tag{4.72}$$

and

$$\hat{\beta}_{\Gamma_4} = (X'X)^+_{\Gamma_4} X'\underline{y} = \frac{1}{12}\begin{bmatrix} 0 \\ 0 \\ 10 \\ 20 \\ -40 \end{bmatrix}. \tag{4.73}$$

Since

$$T_4'X'XT_3 = [\ 0 \quad 0\ ], \tag{4.74}$$

it may be verified that $\hat{\beta}_{\Gamma_3} + \hat{\beta}_{\Gamma_4}$ is a LSS to the normal equations (4.34).

●

4.7 Minimum Norm LSS's

It is known that of all the LSS's of the normal equations there exists a unique LSS having the smallest norm. Since the difference between any two LSS's is in the null space of X, if $\Omega = \mathrm{Row}(X)$ then for any LSS $\hat{\beta}_\Gamma$, $\hat{\beta}_\Omega$ and $\hat{\beta}_\Gamma - \hat{\beta}_\Omega$ are orthogonal. Consequently,

$$\|\hat{\beta}_\Gamma\|^2 = \|\hat{\beta}_\Omega\|^2 + \|\hat{\beta}_\Gamma - \hat{\beta}_\Omega\|^2 \geq \|\hat{\beta}_\Omega\|^2 \tag{4.75}$$

demonstrating that $\hat{\beta}_\Omega$ is the unique LSS having the smallest Euclidean norm.

It is known that the LSS with the smallest N-norm may be expressed as $\hat{\beta} = G'y$ where G is the unique {1,2,3}-inverse of X satisfying $(NGX)' = NGX$. Expressions for G are known. The following may be found on page 123 of Ben-Israel and Greville (1974):

$$G = N^{-1}F'(E'XN^{-1}F')^{-1}E', \tag{4.76}$$

where $X = EF'$ is a full rank factorization. Thus G is a {1,2,3}-inverse of X with $\mathrm{Col}(G) = \Psi = N^{-1}(\Omega)$, where $\Omega = \mathrm{Row}(X)$. Consequently G may be rewritten as

$$G = (X'X)^{+}_{\Psi}X'. \tag{4.77}$$

The LSS to the normal equations with the minimum N-norm is therefore $\hat{\beta}_{\Psi}$.

For an arbitrary space Γ, does there exist a unique LSS satisfying $\hat{\beta} \subset \Gamma$ which has the smallest N-norm and, if so, what is its expression in terms of a Bott-Duffin Inverse?

Recall from Section 4.5 that for any subspace Γ, if $\Gamma \cap \mathrm{Null}(X) = \{0\}$ then the LSS satisfying $\hat{\beta} \in \Gamma$ is unique and consequently it has the smallest N-norm satisfying the constraint.

If $\mathcal{N} = \Gamma \cap \mathrm{Null}(X) \neq \{0\}$ there is no unique LSS satisfying $\hat{\beta} \in \Gamma$. In a manner similar to (4.75) it is easily shown that $\hat{\beta}_{\Psi}$, where $\Psi = \Gamma \cap \mathcal{N}^{\perp}$, has the smallest Euclidean norm of all LSS's satisfying the constraint. However, for an arbitrary positive definite matrix N, which LSS has the smallest N-norm? This problem was considered by Rao and Mitra [(1973), p. 483]. When $\Gamma = \mathrm{Col}(J)$ they called matrices A such that $\hat{\beta} = Ay$ is the minimum N-norm LSS satisfying $\hat{\beta} \in \Gamma$ the "minimum J-restricted least squares" inverse of X. They denoted such inverses as $X^{+}_{(J)}N$. In that paper, they established the necessary and sufficient conditions for $A = X^{+}_{(J)}N$, when $N = I$. No general solution for $X^{+}_{(J)}N$ was offered when $N \neq I$ or $\mathrm{Dim}(\Gamma) < \mathrm{Rank}(X)$.

In the next lemma, a Bott-Duffin expression for minimum N-norm constrained LSS's is presented which holds when $\mathrm{Dim}(\Gamma) < \mathrm{Rank}(X)$. The expression involves only identifying the constraint satisfied by the LSS.

Lemma 4.4: If $\Gamma \not\subset \mathrm{Null}(X)$ and $\mathcal{N} = \Gamma \cap \mathrm{Null}(X)$, then the LSS satisfying $\hat{\beta} \in \Gamma$ with the smallest N-norm is given by $\hat{\beta}_{\Psi}$, where $\Psi = \Gamma \cap N^{-1}(\mathcal{N}^{\perp})$. Ψ is the subspace of Γ which is N-orthogonal to $\mathcal{N}$.

Proof:

1. If $\mathcal{N} = \{\underline{0}\}$ then $\Psi = \Gamma$.
2. If $\mathcal{N} \neq \{\underline{0}\}$, let $N = BB'$ be a full rank factorization. Then

$$\min_{\underline{\beta} \varepsilon \Gamma} \left\| \underline{y} - X\underline{\beta} \right\|^2 = \min_{\underline{\gamma} \varepsilon B'(\Gamma)} \left\| \underline{y} - X(B')^{-1}\underline{\gamma} \right\|^2 \tag{4.78}$$

and
$$\left\| \hat{\underline{\beta}} \right\|_N^2 = \left\| B'\hat{\underline{\beta}} \right\|^2. \tag{4.79}$$

Consequently $\hat{\underline{\beta}}$ is a minimum N-norm constrained LSS for $\underline{\beta} \varepsilon \Gamma$ if and only if $B'\hat{\underline{\beta}}$ is a minimum norm LSS for $\underline{\gamma} \varepsilon B'(\Gamma)$.

Let
$$\mathcal{N}_* = \left\{ B'\underline{\beta} \mid \underline{\beta} \varepsilon \Gamma,\ X(B')^{-1}B'\underline{\beta} = \underline{0} \right\} \tag{4.80}$$

and
$$\Psi_* = \left\{ B'\underline{\beta} \mid \underline{\beta} \varepsilon \Gamma,\ B'\underline{\beta} \varepsilon [B'(\mathcal{N})]^{\perp} \right\}. \tag{4.81}$$

Then as shown above

$$B'\hat{\underline{\beta}} = [B^{-1}X'X(B')^{-1}]^{+}_{\Psi_*} B^{-1}X'\underline{y}\ . \tag{4.82}$$

Consequently,

$$\hat{\underline{\beta}} = (B')^{-1}[B^{-1}X'X(B')^{-1}]^{+}_{\Psi_*} B^{-1}X'\underline{y} \tag{4.83}$$

$$= (X'X)^{+}_{(B')^{-1}(\Psi_*)} X'\underline{y}\ . \tag{4.84}$$

However,

$$(B')^{-1}(\Psi_*) = \left\{ \underline{\beta} \varepsilon \Gamma \mid B'\underline{\beta} \varepsilon [B'(\mathcal{N})]^{\perp} \right\} \tag{4.85}$$

$$= \left\{ \underline{\beta} \varepsilon \Gamma \mid N\underline{\beta} \varepsilon \mathcal{N}^{\perp} \right\} \tag{4.86}$$

$$= \Gamma \cap N^{-1}(\mathcal{N}^{\perp}). \tag{4.87}$$

●

As a specific application of the above lemma, the minimum N-norm LSS satisfying the estimable constraints $L'\underline{\beta} = \underline{0}$ is given below.

Corollary 4.4.1: The unique LSS with the minimum N-norm satisfying $L'\underline{\beta} = \underline{0}$ is given by

$$\hat{\underline{\beta}}_{\Psi} = (X'X)^{+}_{\Psi} X'\underline{y}, \tag{4.88}$$

where $\Psi = N^{-1}\left\{ \underline{\omega} \varepsilon \Omega \mid L'N^{-1}\underline{\omega} = \underline{0} \right\}$, $\Omega =$ Row(X).

Proof: Follows from Theorem 4.4 since $\mathcal{N} =$ Null(X) and $\mathcal{N}^{\perp} = \Omega$.

●

4.8 A General Theorem On Constrained LSS's

The above theorems and corollaries lead to the following theorem concerning the expression of LSS's satisfying nonhomogeneous constraints, which minimize the weighted sum of squares (4.1).

Theorem 4.5: $\hat{\underline{\beta}}$ is a LSS minimizing

$$\|\underline{y} - x\underline{\beta}\|_W^2 \tag{4.89}$$

subject to the constraint $(\underline{\beta} - \underline{t}) \varepsilon \Gamma$ if and only if

$$\hat{\underline{\beta}}_\Psi^* = (X'WX)_\Psi^+ X'W(\underline{y} - X\underline{t}) + \underline{t} \tag{4.90}$$

where $\Psi \subset \Gamma$, $\wp_\Omega(\Psi) = \wp_\Omega(\Gamma)$, $\Omega = \text{Row}(X)$, and $\wp_\Omega$ is the Euclidean projection onto Ω.

Proof: Let $W = TT'$ be a full rank factorization, then

$$\min_{(\underline{\beta}-\underline{t})\varepsilon\Gamma} \|\underline{y} - X\underline{\beta}\|_W^2 = \min_{\underline{\beta}\varepsilon\Gamma} \|\underline{y} - X(\underline{\beta} + \underline{t})\|_W^2 \tag{4.91}$$

$$= \min_{\underline{\beta}\varepsilon\Gamma} \|T'\underline{y} - T'X(\underline{\beta} + \underline{t})\|^2 \tag{4.92}$$

$$= \min_{\underline{\beta}\varepsilon\Gamma} \|T'(\underline{y} - X\underline{t}) - T'X\underline{\beta}\|^2. \tag{4.93}$$

From Theorem 4.3, $\hat{\underline{\beta}}$ satisfies (4.93) if and only if

$$\hat{\underline{\beta}} = (X'TT'X)_\Psi^+ X'TT'(\underline{y} - X\underline{t}) \tag{4.94}$$

for some Ψ satisfying the conditions of the theorem.

●

The minimum N-norm LSS satisfying a set of nonhomogeneous constraints is characterized in the following corollary.

Corollary 4.5.1: $\hat{\underline{\beta}}$ is the unique minimum N-norm constrained LSS minimizing (4.88) and satisfying $(\hat{\underline{\beta}} - \underline{t}) \varepsilon \Gamma$ if and only if

$$\hat{\underline{\beta}} = \hat{\underline{\beta}}_\Psi^*, \tag{4.95}$$

$\Psi = \Gamma \cap N^{-1}(\mathcal{N}^\perp)$, where $\mathcal{N} = \Gamma \cap \text{Null}(X'WX)$.

●

4.9 Residual Sum of Squares And Their Difference

Up to this point, the concern has been in characterizing LSS's and establishing Bott-Duffin Inverses as the minimum sufficent set of matrices for generating LSS's. In this section, attention is turned to the expression, in terms of Bott-Duffin Inverses, of the residual sum of squares. This leads to a simplified expression for the difference between the residual sum of squares associated with a pair of nested constraints.

Lemma 4.6: For spaces $\Gamma_1 \subset \Gamma_2$ disjoint from Null(X),

$$Q_1 = \min_{\underline{\beta}\in\Gamma_1} \|\underline{y} - X\underline{\beta}\|^2 = \underline{y}'\underline{y} - \hat{\underline{\beta}}'_{\Gamma_1} X'X \hat{\underline{\beta}}_{\Gamma_1} \quad (4.96)$$

$$Q_2 = \min_{\underline{\beta}\in\Gamma_2} \|\underline{y} - X\underline{\beta}\|^2 = \underline{y}'\underline{y} - \hat{\underline{\beta}}'_{\Gamma_2} X'X \hat{\underline{\beta}}_{\Gamma_2} \quad (4.97)$$

and

$$Q_1 - Q_2 = \hat{\underline{\beta}}'_{\Gamma_{2*1}} X'X \hat{\underline{\beta}}_{\Gamma_{2*1}} \quad (4.98)$$

where $\Gamma_{2*1} = \left\{ \underline{\gamma}_2 \varepsilon \Gamma_2 \mid \underline{\gamma}_2' X'X \underline{\gamma}_1 = 0 \text{ for all } \underline{\gamma}_1 \varepsilon \Gamma_1 \right\}$.

Proof: From Corollary 4.3.3, for i = 1, 2, 2∗1

$$X\hat{\underline{\beta}}_{\Gamma_i} = X(X'X)^+_{\Gamma_i} X'X \hat{\underline{\beta}}_{\Gamma_i} , \quad (4.99)$$

which implies

$$\underline{y}'X\hat{\underline{\beta}}_{\Gamma_i} = \hat{\underline{\beta}}'_{\Gamma_i} X'X\hat{\underline{\beta}}_{\Gamma_i} . \quad (4.100)$$

(4.99) and (4.100) follow immediately. From Corollary 4.3.4

$$\hat{\underline{\beta}}_{\Gamma_2} = \hat{\underline{\beta}}_{\Gamma_1} + \hat{\underline{\beta}}_{\Gamma_{2*1}} \quad (4.101)$$

and by definition of Γ_{2*1}

$$0 = \hat{\underline{\beta}}'_{\Gamma_{2*1}} X'X \hat{\underline{\beta}}_{\Gamma_1}. \quad (4.102)$$

Together (4.101) and (4.102) imply

$$\hat{\underline{\beta}}'_{\Gamma_2} X'X \hat{\underline{\beta}}_{\Gamma_2} = \hat{\underline{\beta}}'_{\Gamma_1} X'X \hat{\underline{\beta}}_{\Gamma_1} + \hat{\underline{\beta}}'_{\Gamma_{2*1}} X'X \hat{\underline{\beta}}_{\Gamma_{2*1}} \quad (4.103)$$

The expression (4.98) follows immediately.

•

Expressions (4.96) and (4.97) are similar to the well known expressions in linear model analysis. The advantage of expression (4.98) is in providing a straightforward proof of the correspondence between the reduction in residual sum of squares approach and the "inverse of the inverse" approach to hypothesis testing in linear models. This will be discussed in the next chapter.

4.10 Computing Constrained LSS's and Residual Sum of Squares

As discussed in Chapter II, the G2SWEEP operator described by Goodnight (1979) provides an easy way for computing {2}-inverses. This algorithm may be extended to obtain a LSS and the corresponding residual sum of squares for any constraint.

To obtain a LSS corresponding to the constraint $\hat{\underline{\beta}} \varepsilon \Gamma$, first choose matrix G such that Col(G) = Γ and then form the following matrix:

$$\begin{bmatrix} G'X'XG & G'X'\underline{y} \\ \underline{y}'XG & \underline{y}'\underline{y} \end{bmatrix} \tag{4.104}$$

sweeping the upper left corner matrix yields

$$H = \begin{bmatrix} \left(G'X'XG\right)^{+}_{1,2} & \left(G'X'XG\right)^{+}_{1,2}G'X'\underline{y} \\ -\underline{y}'XG\left(G'X'XG\right)^{+}_{1,2} & \underline{y}'\underline{y} - \underline{y}'XG\left(G'X'XG\right)^{+}_{1,2}G'X'\underline{y} \end{bmatrix}. \tag{4.105}$$

Premultiplying H by

$$A = \begin{bmatrix} G & 0 \\ 0 & I \end{bmatrix} \tag{4.106}$$

and postmultiplying by A′ gives

$$AHA' = \begin{bmatrix} G\left(G'X'XG\right)^{+}_{1,2}G' & G\left(G'X'XG\right)^{+}_{1,2}G'X'\underline{y} \\ -\underline{y}'XG\left(G'X'XG\right)^{+}_{1,2}G' & \underline{y}'\underline{y} - \underline{y}'XG\left(G'X'XG\right)^{+}_{1,2}G'X'\underline{y} \end{bmatrix} \tag{4.107}$$

$$= \begin{bmatrix} (X'X)^{+}_{\Psi} & \hat{\underline{\beta}}_{\Psi} \\ -\hat{\underline{\beta}}_{\Psi} & \left\|\underline{y} - X\hat{\underline{\beta}}_{\Psi}\right\|^{2} \end{bmatrix}. \tag{4.108}$$

If $G'X'XG$ is nonsingular, then in (4.108) $\Psi = \Gamma$ and $\hat{\underline{\beta}}_{\Psi}$ is the unique LSS satisfying $\hat{\underline{\beta}} \in \Gamma$. Otherwise Ψ is a proper subspace of Γ with $\wp_{\Omega}\Psi = \wp_{\Omega}\Gamma$ and $\Psi \cap \text{Null}(X) = \{\underline{0}\}$. The above algorithm does not depend upon a factorization of X or on solving for its eigenvectors, methods suggested by Golub and Van Loan [(1983), pp. 405-412].

For a set of constraints $H'\underline{\beta} = \underline{0}$, where H' is of full row rank r, the algorithm requires finding a basis matrix for the space Γ orthogonal to Col(H). One way to find such a basis is to perform a Gram-Schmidt orthogonalization from left to right on the columns of

$$\begin{bmatrix} H & I \end{bmatrix}. \tag{4.109}$$

The resulting matix will be of the form

$$\begin{bmatrix} H_* & I_* \end{bmatrix} \tag{4.110}$$

where H_* is an orthonormal basis of Col(H) and the nonzero columns of I_* will span the space orthogonal to Col(H).

Alternatively, by the application of repeated row operations on H', it may be assumed without loss of generality that

$$H' = \begin{bmatrix} I & H_*' \end{bmatrix}, \tag{4.111}$$

then the columns of

$$\begin{bmatrix} -H_*' \\ I \end{bmatrix} \tag{4.112}$$

will also span the space orthogonal to Col(H).

CHAPTER V
{2}-INVERSES IN LINEAR MODELS

5.0 Introduction

For an experiment designed with an equal number of replications in each cell, there is little controversy concerning the sum of squares to be used in testing the various effects. However, in reality most data are imbalanced and there is no unanimity of opinion as to the sum of squares appropriate for testing the various effects.

As noted by Goodnight (1978), following Kutner's (1974) article, statisticians took sides and split into the R and R^* camps. The R camp uses the overparameterized model and computes the sum of squares for an effect, $\underline{\beta}$, as the sum of squares due to $\underline{\beta}$ after adjusting for all effects other than those of higher order which contain $\underline{\beta}$. On the other hand, the R^* camp reparameterizes to a full rank model using the "usual" constraints and then computes the sum of squares for an effect after adjusting for all other effects.

In the usual context, the R and the R^* approaches are equivalent respectively to the method of fitting constants and the method of weighted squares of means, both proposed by Yates (1934).

Hocking and Speed (1975), and Speed and Hocking (1976) demonstrated that the hypothesis tested by the R approach is not identical to that tested by the R^* approach. In addition, the hypothesis tested by the R^* approach is not easily expressed in terms of the parameters of the original overparameterized model. The authors advocated the benefit of expressing hypotheses in terms of the cell means, an approach they claim avoids the ambiguity of poorly defined hypotheses.

Following this work Speed, Hocking and Hackney (1978) reviewed a number of methods for analyzing models with unbalanced data. They expressed the hypotheses tested by these methods in terms of the cell means. In particular, they considered the R and R^* approaches and pointed out that for the two-way interaction model, the hypothesis associated with the R approach depends upon the cell sizes while that associated with the R^* approach does not. They suggested that a hypothesis depending on particular cell frequencies is more difficult to justify than one independent of the cell frequencies.

Searle, Speed and Henderson (1981) described the most common techniques of calculating sums of squares found in the literature. They explained the relationships between these methods and illustrated them in terms of the two-way classification model with interaction. They also advocated the utility of the cell means in determining the hypotheses tested by the various methods.

Interest in the differences between the R and R^* approaches in computing sums of squares and their corresponding hypotheses is more than strictly academic. The premier statistical package today, the Statistical Analysis System (i.e. the SAS System) includes in its general linear model procedure, PROC GLM, both the R and the R^* approaches as their Type II and Type III sums of squares, respectively. The most recent SAS Users' Manual (1985) contains a lengthy section devoted to describing SAS' approach to linear model analysis which overall depends upon the G2SWEEP operator described by Goodnight (1979). The computation of the SAS Type I, II, III and IV sums of squares is based upon various sets of estimable contrasts constructed according to different algorithms. However, these algorithms provide little insight into the nature of the hypotheses they test.

Numerous authors have sought to end this confusion by providing examples illustrating specific concepts, most notably Speed and Hocking (1980), and Freund and Littell (1981).

The need to clearly define the hypotheses tested by the various approaches has encouraged researchers to develop algorithms for translating hypotheses from the terms of the original model into contrasts of the cell means. These include papers by Hsuan (1982), Smith and Murray (1983), Schmoyer (1984), and Tandon and Lustick (1985).

It was demonstrated in Chapter IV that Bott-Duffin Inverses provide a computationally tractable way to generate constrained LSS's to normal equations. This work suggests that a {2}-inverse point of view of linear models may offer an advantage in providing a consistent theory unifying the R, R^*, and the cell means approaches.

The aim of this chapter is to demonstrate that Bott-Duffin Inverses do indeed provide such an advantage. This will be done in the next seven sections outlined below:

5.1 In this section, four alternative formulations of a linear model are given: the Unconstrained Model, the Constrained Model where the parameters of the model are assumed to satisfy a set of known constraints, the Reparameterized Reduced Model where the constraints are used to reparameterize the model and reduce the number of parameters, and the Cell Means Model of Hocking and Speed (1975) in which a set of constraints is forced on the cell means to assure the equivalence of this model to some parametric one.

5.2 In the three lemmas of this section, the relationships existing among the LSS's of the parameters of the various models in Section 5.1 are established. The expressions of these LSS's in terms of the sample cell means are also given.

5.3 Five approaches to formulating hypotheses are given: one each in terms of the above four models and the fifth in terms of the cell sample means. In Lemma 5.5 the "inverse of the inverse" approach for computing the appropriate numerator sum of squares in the unconstrained model is extended to hypothesis testing in a constrained model.

5.4 Theorem 5.6 establishes sufficient conditions on the testable hypotheses in each of the above five analytical approaches which assures the equality of the associated numerator sums of squares. This leads to a series of algorithms for defining the equivalent hypotheses tested in each analytical approach.

5.5 Lemma 5.7 gives a sufficient set of conditions under which the hypotheses tested by the five analytical approaches are invariant to the cell sizes.

5.6 In this section, the results of this chapter are applied to determine the hypotheses tested by the SAS Type I and II Sums of Squares.

5.7 In this section, the results of this chapter are applied to hypotheses tested in full rank models reparameterized from the original by assuming the "usual" constraints, an approach which leads to the SAS Type III hypotheses.

5.1 The Models

In most analytical situations where a linear model is appropriate there are several alternative formulations. In this section, the notations are established for the various models which will be considered in this chapter.

Definition 5.1: The Unconstrained Model is the fixed effects model

$$\underline{y} = X\underline{\beta}+\underline{\epsilon} \tag{5.1}$$

where $\underline{y}$ is an $n{\times}1$ vector of random variables,

X is an $n{\times}p$ design matrix of known constants of rank $r \leq p$,

$\underline{\beta}$ is a $p{\times}1$ vector of unknown parameters, and

$\underline{\epsilon}$ is an $n{\times}1$ vector of random variables distributed as $\mathcal{N}(\underline{0},\sigma^2 I)$.

●

For this model it is convenient to make the following definition:

Definition 5.2: Write $X = WX_*$, where the cell incidence matrix, W, is the $n{\times}c$ full column rank incidence matrix linking each observation to a particular cell, and the cell model matrix, X_*, is formed from X by deleting duplicate rows. If X_* is full row rank, then the model (5.1) is saturated.

●

As popularized by Hocking and Speed (1975), model (5.1) may also be expressed in terms of the cell means in what they call the Cell Means Model.

Definition 5.3: The Cell Means Model is the model

$$\underline{y} = W\underline{\mu}+\underline{\epsilon} \tag{5.2}$$

with $\underline{\mu}$ satisfying

$$\underline{0} = G'\underline{\mu}\ , \tag{5.3}$$

where W is the cell incidence matrix,

$\underline{\mu}$ is a $c{\times}1$ parameter vector of cell means, and

G is a $c{\times}(c-r)$ known full column rank matrix expressing a linear constraint on the cell means.

●

$\underline{\beta}$ and $\underline{\mu}$ in the above two models are related through the relationship

$$X_* \underline{\beta} = \underline{\mu} \tag{5.4}$$

and G statisfies

$$\mathrm{Col}(G) = \mathrm{Null}(X_*'). \tag{5.5}$$

From (5.5) it is clear that the model (5.1) is saturated if and only if G is identically a null matrix. Thus, when (5.1) is not saturated the equivalent cell means model is constrained. As an example, the usual two-way cross classification model without interaction

$$y_{ijk} = \underline{\mu}+\alpha_i+\beta_j+\epsilon_{ijk} \tag{5.6}$$

is equivalent to the cell means model where the cell means satisfy the constraint

$$0 = m_{ij} - m_{i'j} - m_{ij'} + m_{i'j'}. \tag{5.7}$$

The latter model is equivalent to the two-way cross classification model with interaction

$$y_{ijk} = \mu+\alpha_i+\beta_j+\gamma_{ij}+\epsilon_{ijk} \tag{5.8}$$

satisfying the constraints

$$0 = \gamma_{ij} - \gamma_{i'j} - \gamma_{ij'}+\gamma_{i'j'}. \tag{5.9}$$

In practice, unsaturated models are not thought of as a restriction of some larger model. Consider the classical balanced incomplete block design. It is not readily apparent what the equivalent restricted saturated model is nor what the appropriate restrictions on the cell means might be.

Consequently, from this point on, a model will be considered as constrained only if the parameters of the model, saturated or not, are assumed to satisfy a set of known linear restrictions. This leads to the following definition:

Definition 5.4: Model (5.1) is a Constrained Model if $\underline{\beta}$ is assumed, as part of the model, to satisfy the consistent constraints $K'\underline{\beta} = \underline{k}$, or equivalently $\underline{\beta} - \underline{t} \in G$, where K' is an $s \times p$ full row rank set of estimable or nonestimable constraints.

•

The above definition includes the case where p-r linearly independent nonestimable homogeneous constraints are imposed on model (5.1) to uniquely solve the normal equations. In this circumstance, model (5.1) can be reparameterized to one of full rank. With this as motivation, the following definition is made.

Definition 5.5: Corresponding to the homogeneous constaints $K'\underline{\beta} = \underline{0}$, in the constrained model, the model

$$\underline{y} = X\tilde{K}\underline{\tilde{\beta}}+\underline{\epsilon} \tag{5.10}$$

is the reparameterized reduced model, where $\tilde{K}$ is a $p{\times}(p\text{-}s)$ full column rank matrix such that

$$\tilde{K}\underline{\tilde{\beta}} = \underline{\beta} \tag{5.11}$$

and

$$K'\tilde{K} = 0. \tag{5.12}$$

When $X\tilde{K}$ is full column rank then (5.10) is a full rank reparameterization.

•

Several types of constraints commonly appear in the literature. For example, the following may be found in Searle, Speed and Henderson (1981). The Σ-constraints, or "usual" constraints, which in the two-way model are

$$y_{ijk} = \mu+\alpha_i+\beta_j+\gamma_{ij}+\epsilon_{ijk}, \tag{5.13}$$

are: $\alpha_+=0$, $\beta_+=0$, $\gamma_{i+}=0$ and $\gamma_{+j} = 0$ for all i, j.

The 0-constraints, where some elements of the parameter vector are set equal to zero, For example:

$$\alpha_1 = 0,\ \gamma_{i1} = 0,\ \beta_1 = 0,\ \gamma_{1j} = 0 \quad \text{for each i, j.} \tag{5.14}$$

The W-constraints, requiring a weighted sum of the parameters to be zero, for example:

$$\begin{aligned} &\sum_i n_{i.}\alpha_i = 0, \quad \sum_j n_{ij}\gamma_{ij} = 0, \text{ for each i,} \\ &\sum_j n_{j.}\beta_j = 0, \quad \sum_i n_{ij}\gamma_{ij} = 0, \text{ for each j,} \end{aligned} \tag{5.15}$$

where n_{ij} (>0) is the size of the ij-th cell.

Although not commonly done in the literature, the reduced models, such as those associated with the reduction sum of squares, may also be expressed in terms of model (5.10). For example, suppose X is partitioned as $\begin{bmatrix} X_1 & X_2 \end{bmatrix}$ and $\underline{\beta}'$ is partitioned conformably as $\begin{bmatrix} \underline{\beta_1} & \underline{\beta_2} \end{bmatrix}$ then the reduced model

$$\underline{y} = X_1\underline{\beta_1}+\underline{\epsilon} \tag{5.16}$$

is equivalent to (5.10) with $\tilde{K}' = \begin{bmatrix} I & 0 \end{bmatrix}$ and $\underline{\tilde{\beta}} = \underline{\beta_1}$.

5.2 The Distribution and Relationships Among The LSS's For The Parameters in Various Models

In the previous section, four models were identified: the unconstrained model, its equivalent cell means model, the constrained model, and the corresponding reparameterized reduced model. In this section the aim is to apply the results of the previous chapter to finding the LSS's for the parameters in these various models. This leads to an expose′ of the simple relationships existing among these LSS's.

Lemma 5.1: For the model (5.1), and spaces Γ_1 and Γ_2 disjoint from Null(X) then for i = 1, 2

$$\hat{\underline{\beta}}_{\Gamma_i} = (X'X)^+_{\Gamma_i} X'\underline{y} = (X'X)^+_{\Gamma_i} X'W\underline{z} = (X'X)^+_{\Gamma_i} X'X\hat{\underline{\beta}}\,, \tag{5.17}$$

$$\hat{\underline{\beta}}_{\Gamma_i} \sim \mathcal{N}\left[(X'X)^+_{\Gamma_i} X'X\underline{\beta}\,,(X'X)^+_{\Gamma_i}\sigma^2\right] \tag{5.18}$$

and

$$\mathrm{Cov}\left[\hat{\underline{\beta}}_{\Gamma_1},\hat{\underline{\beta}}_{\Gamma_2}\right] = (X'X)^+_{\Gamma_1} X'X(X'X)^+_{\Gamma_2}\sigma^2, \tag{5.19}$$

where $\underline{z}$ is the vector of sample cell means and $\hat{\underline{\beta}}$ is any LSS to the normal equations $X'X\underline{\beta} = X'\underline{y}$. Thus $\hat{\underline{\beta}}_{\Gamma_1}$ and $\hat{\underline{\beta}}_{\Gamma_2}$ are independently distributed if and only if Γ_1 and Γ_2 are (X′X)-orthogonal.

Proof: Follows directly from the definition of the $\hat{\underline{\beta}}_{\Gamma_i}$ and the equality

$$X'W\underline{z} = X_*'\, W'W(W'W)^{-1}W'\underline{y} = X_*'\, W'\underline{y} = X'\underline{y}\,. \tag{5.20}$$

●

The following is an interesting corollary to Lemma 5.1.

Corollary 5.1.1: Let $\Gamma_1 \subset \Gamma_2$ be subspaces disjoint from Null(X) then

$$\hat{\underline{\beta}}_{\Gamma_2} = \hat{\underline{\beta}}_{\Gamma_1} + \hat{\underline{\beta}}_{\Gamma_{2*1}} \tag{5.21}$$

where $\Gamma_{2*1} = \left\{\underline{\gamma}_2 \in \Gamma_2 \mid \underline{\gamma}_2 X'X\underline{\gamma}_1 = 0 \text{ for all } \underline{\gamma}_1 \in \Gamma_1\right\}$, and $\hat{\underline{\beta}}_{\Gamma_1}$ and $\hat{\underline{\beta}}_{\Gamma_{2*1}}$ are independently distributed.

Proof: Follows from Corollary 4.3.4 and the definition of Γ_{2*1}.

●

Theorem 4.3 may also be applied to the cell means model to prove the following lemma.

Lemma 5.2: The LSS of $\underline{\mu}$ in the Cell Means Model is uniquely given by

$$\hat{\underline{\mu}}_{\Phi} = D^+_{\Phi}W'\underline{y} = D^+_{\Phi}D\underline{z} = X_*\hat{\underline{\beta}}\,, \tag{5.22}$$

such that $$\hat{\underline{\mu}}_{\Phi} \sim \mathcal{N}\left[D_{\Phi}^{+}D\underline{\mu}\ ,\ D_{\Phi}^{+}\sigma^2\right],\tag{5.23}$$

where $\Phi = \mathrm{Col}(X_*)$,

$D = W'W$ is the diagonal matrix of cell sizes,

$\hat{\underline{\beta}}$ is any LSS to the normal equations $X'X\underline{\beta} = X'\underline{y}$, and

$\underline{z}$ is the vector of sample cell means.

Furthermore, for any constrained LSS of $\underline{\beta}$

$$\hat{\underline{\beta}}_{\Gamma} = (X'X)_{\Gamma}^{+}X'W\hat{\underline{\mu}}_{\Phi}.\tag{5.24}$$

Proof: Trivially $D\underline{z} = W'\underline{y}$. In addition for any g-inverse $(X'X)^{-}$,

$$D_{\Phi}^{+}W'\underline{y} = (W'W)_{\Phi}^{+}W'\underline{y} = X_*(X_*' W'WX_*)^{-}X_*' W'\underline{y} = X_*(X'X)^{-}X'\underline{y}\ ,\tag{5.25}$$

thus implying (5.22). For the converse

$$\hat{\underline{\beta}}_{\Gamma} = (X'X)_{\Gamma}^{+}X'X\hat{\underline{\beta}} = (X'X)_{\Gamma}^{+}X'WX_*\hat{\underline{\beta}} = (X'X)_{\Gamma}^{+}X'W\hat{\underline{\mu}}_{\Phi}\ .\tag{5.26}$$

●

In the above lemma, if the model (5.1) is saturated then $D_{\Phi}^{+} = D^{-1}$ and $\hat{\underline{\mu}} = \underline{z}$.

Lemma 5.3: For model (5.10)

1. If $\hat{\hat{\underline{\beta}}}_{\Psi}$ is a LSS for the normal equations of model (5.10), then

$$\hat{\underline{\beta}}_{\tilde{K}(\Psi)} = \tilde{K}\hat{\hat{\underline{\beta}}}_{\Psi}\tag{5.27}$$

is a LSS for $\underline{\beta}$ in model (5.1) satisfying $K'\underline{\beta} = \underline{0}$.

2. If $\hat{\underline{\beta}}_{\Gamma}$ is a LSS for $\underline{\beta}$ in model (5.1) satisfying $K'\underline{\beta} = \underline{0}$, then

$$\hat{\hat{\underline{\beta}}}_{\Upsilon} = \left(\tilde{K}'\tilde{K}\right)^{-1}\tilde{K}'\hat{\underline{\beta}}_{\Gamma}\tag{5.28}$$

is a LSS of the normal equations of model (5.10) where $\Upsilon = \left(\tilde{K}'\tilde{K}\right)^{-1}\tilde{K}'(\Gamma)$.

Proof:

1. The following equality follows from Corollary 2.8.5

$$\left(X'X\right)_{\tilde{K}(\Psi)}^{+} = \tilde{K}\left(\tilde{K}'X'X\tilde{K}\right)_{\Psi}^{+}\tilde{K}'.\tag{5.29}$$

2. Since $\Gamma \subset \mathrm{Col}(\tilde{K})$ for an appropriate g-inverse of $\tilde{K}'X'X\tilde{K}$

$$\hat{\underline{\beta}}_{\Gamma} = \tilde{K}\left(\tilde{K}'X'X\tilde{K}\right)^{-}\tilde{K}'X'\underline{y}\ .\tag{5.30}$$

●

5.3 Hypothesis Testing in Linear Models

In this section the concern will be with the expression of the numerator sum of squares for testing a null hypothesis formulated in terms of the parameters of the four models defined in section 5.1. While different formulations of a hypothesis are possible, the present discussion will be limited to those expressible in terms of estimable functions of the parameters of the particular model.

Definition 5.6: The hypothesis

$$H_0\colon L'\underline{\beta} = \underline{m} \tag{5.31}$$

is testable if (5.31) is a consistent set of equations and each row of L' is an estimable function.

•

It is well known that if $\underline{\lambda}'\underline{\beta}$ is estimable then for some $\underline{t}'$, $\underline{\lambda}'\underline{\beta} = \mathcal{E}(\underline{t}'\underline{y})$. Furthermore, although there is no unique LSS to the normal equations, the estimable functions of the LSS's are unique. The Gauss-Markoff Theorem stated below is the fundamental result concerning estimable functions.

Theorem 5.4: The Best Linear Unbiased Estimate (BLUE) of the estimable function $\underline{\lambda}'\underline{\beta}$ is the unique estimate $\underline{\lambda}'\hat{\underline{\beta}}_\Gamma$, where $\underline{\beta}_\Gamma$ is any solution to the normal equations.

•

If $L'\underline{\beta}$ is testable, then the numerator sum of squares for testing the hypothesis Ho: $L'\underline{\beta} = \underline{m}$ in the unconstrained model is given by

$$SS[H_0\colon L'\underline{\beta} = \underline{m}\,] = \underset{L'\underline{\beta}=\underline{m}}{\text{Min}} \left\| \underline{y} - X\underline{\beta} \right\|^2 - \underset{\underline{\beta}}{\text{Min}} \left\| \underline{y} - X\underline{\beta} \right\|^2 \tag{5.32}$$

and in the constrained model by

$$SS[H_0\colon L'\underline{\beta} = \underline{m} \mid K'\underline{\beta} = \underline{0}] = \underset{\substack{K'\underline{\beta}=\underline{0} \\ L'\underline{\beta}=\underline{m}}}{\text{Min}} \left\| \underline{y} - X\underline{\beta} \right\|^2 - \underset{K'\underline{\beta}=\underline{0}}{\text{Min}} \left\| \underline{y} - X\underline{\beta} \right\|^2. \tag{5.33}$$

It is well known, e.g. Searle [(1971), p. 190] that in the unconstrained model

$$SS[H_0\colon L'\underline{\beta} = \underline{m}\,] = \left(L'\hat{\underline{\beta}}_\Gamma - \underline{m}\,\right)\left(L'(X'X)^+_\Gamma L\right)^{-1}\left(L'\hat{\underline{\beta}}_\Gamma - \underline{m}\,\right) \tag{5.34}$$

where $\hat{\underline{\beta}}_\Gamma$ is any LSS to the normal equations. The above expression is referred to in the literature as the "inverse of the inverse" approach to hypothesis testing. It is a special case of the following lemma.'

Lemma 5.5: In the linear model (5.1) where the parameters are assumed to satisfy the constraints $K'\underline{\beta} = \underline{0}$, then assuming Col(L) is linearly independent of Col(K)

$$SS\left[H_0\colon L'\underline{\beta} = \underline{m} \mid K'\underline{\beta} = \underline{0}\right] = \left(L'\hat{\underline{\beta}}_\Gamma - \underline{m}\right)'\left(L'(X'X)^+_\Gamma L\right)^{-1}\left(L'\hat{\underline{\beta}}_\Gamma - \underline{m}\right) \tag{5.35}$$

where $\hat{\underline{\beta}}_\Gamma$ is any LSS satisfying $K'\underline{\beta} = \underline{0}$.

Proof: Let $\Gamma_1 = \left\{\underline{\gamma} \in \Gamma_2 \mid L'\underline{\gamma} = \underline{0}\right\}$, where $\Gamma_2 = \Gamma$, and let Γ_3 be any r-dimensional subspace disjoint from Null(X) such that $\Gamma_2 \subset \Gamma_3$. In addition let

$$\Gamma_{3*1} = \left\{\underline{\gamma}_3 \in \Gamma_3 \mid \underline{\gamma}_3'X'X\underline{\gamma}_1 = 0 \text{ for all } \underline{\gamma}_1 \in \Gamma_1\right\}, \tag{5.36}$$

$$\Gamma_{3*2} = \left\{\underline{\gamma}_3 \in \Gamma_3 \mid \underline{\gamma}_3'X'X\underline{\gamma}_2 = 0 \text{ for all } \underline{\gamma}_2 \in \Gamma_2\right\}, \tag{5.37}$$

and

$$\Gamma_{2*1} = \left\{\underline{\gamma}_2 \in \Gamma_2 \mid \underline{\gamma}_2'X'X\underline{\gamma}_1 = 0 \text{ for all } \underline{\gamma}_1 \in \Gamma_1\right\}. \tag{5.38}$$

For simplicity we will write

$$\hat{\underline{\beta}}_{\Gamma_i} = \hat{\underline{\beta}}_i \text{ and } (X'X)^+_{\Gamma_i} = (X'X)^+_i \tag{5.39}$$

for i = 1, 2, 3, 3*1, 2*1 and 3*2. The following relationships hold:

$$(X'X)^+_{3*1} = (X'X)^+_{3*2} + (X'X)^+_{2*1} \tag{5.40}$$

and

$$(X'X)^+_3 = (X'X)^+_{3*1} + (X'X)^+_1. \tag{5.41}$$

By the conditions of the lemma, there exists a $\underline{t}$ such that $K'\underline{t}$ and $L'\underline{t} = \underline{m}$. In addition, there exists $\underline{\gamma} \in \Gamma_2$ such that $X\underline{t} = X\underline{\gamma}$. Therefore, if Q equals the sum of squares defined by (5.33), then by Theorem 4.5,

$$Q = \left\|\underline{y} - X\hat{\underline{\beta}}_1^*\right\|^2 - \left\|\underline{y} - X\hat{\underline{\beta}}_2\right\|^2, \tag{5.42}$$

where

$$\hat{\underline{\beta}}_1^* = (X'X)^+_1X'(\underline{y} - X\underline{t}) + \underline{t}. \tag{5.43}$$

From the choice of $\underline{t}$, it follows that

$$\begin{aligned} X\hat{\underline{\beta}}_1^* &= X\hat{\underline{\beta}}_1 + X\left[I - (X'X)^+_1X'X\right]\underline{t} && (5.44)\\ &= X\hat{\underline{\beta}}_1 + \left[X(X'X)^+_3X'X - X(X'X)^+_1X'X)\right]\underline{t} && (5.45)\\ &= X\hat{\underline{\beta}}_1 + \left[X(X'X)^+_{3*1}X'X\right]\underline{t} && (5.46)\\ &= X\hat{\underline{\beta}}_1 + \left[X(X'X)^+_{3*2}X'X + X(X'X)^+_{2*1}X'X\right]\underline{t} && (5.47)\\ &= X\hat{\underline{\beta}}_1 + X(X'X)^+_{2*1}X'X\underline{t}. && (5.48)\end{aligned}$$

Furthermore, it follows from Corollary 4.3.3 that

$$X(X'X)^+_{2*1}X'X\hat{\underline{\beta}}_3 = X(X'X)^+_{2*1}X'X\hat{\underline{\beta}}_2 \tag{5.49}$$

and

$$\underline{0} = X(X'X)^+_{2*1}X'X\hat{\underline{\beta}}_1. \tag{5.50}$$

This implies

$$\|\underline{y} - X\hat{\underline{\beta}}_1^*\|^2 = \|\underline{y} - X\hat{\underline{\beta}}_1\|^2 - 2\,\hat{\underline{\beta}}_2' X'X(X'X)_{2*1}^{+}X'X\underline{t} + \underline{t}'X'X(X'X)_{2*1}^{+}X'X\underline{t} \tag{5.51}$$

and

$$Q = \hat{\underline{\beta}}_{2*1}' \; X'X\hat{\underline{\beta}}_{2*1} - 2\,\hat{\underline{\beta}}_2' X'X(X'X)_{2*1}^{+}X'X\underline{t} + \underline{t}'X'X(X'X)_{2*1}^{+}X'X\underline{t} \tag{5.52}$$

$$= \hat{\underline{\beta}}_2' X'X(X'X)_{2*1}^{+}X'X\hat{\underline{\beta}}_2 - 2\,\hat{\underline{\beta}}_2' X'X(X'X)_{2*1}^{+}X'X\underline{t} + \underline{t}'X'X(X'X)_{2*1}^{+}X'X\underline{t} \tag{5.53}$$

$$= \left(\hat{\underline{\beta}}_2 - \underline{t}\right)' X'X(X'X)_{2*1}^{+}X'X\left(\hat{\underline{\beta}}_2 - \underline{t}\right) \tag{5.54}$$

$$= \left(\hat{\underline{\beta}}_2 - \underline{t}\right)'\left[(X'X)_2^{+}\right]_{X'X(\Gamma_{2*1})}^{+}\left(\hat{\underline{\beta}}_2 - \underline{t}\right) \tag{5.55}$$

$$= \left(\hat{\underline{\beta}}_2 - \underline{t}\right)'\left[(X'X)_2^{+}\right]_{\Lambda}^{+}\left(\hat{\underline{\beta}}_2 - \underline{t}\right) \tag{5.56}$$

$$= \left(L'\hat{\underline{\beta}}_2 - \underline{m}\right)'\left(L'(X'X)_2^{+}L\right)^{-1}\left(L'\hat{\underline{\beta}}_2 - \underline{m}\right) . \tag{5.57}$$

●

In addition to the four alternative models discussed in Section 5.1, a hypothesis of interest may also be expressed in terms of the sample cell means, $\underline{z}$, ignoring the constraints, (5.3), imposed by an assumed parameterized model. Nevertheless, regardless of the analytical approach used, Lemma 5.5 provides an efficient expression for the appropriate numerator sum of squares for a testable hypothesis. The expression for these various sums of squares is summarized in the following corollary:

Corollary 5.5.1:

1. For $L'\underline{\beta} = \underline{m}$, a testable hypothesis in the unconstrained model (5.1),

$$SS\left[H_0\!: L'\underline{\beta} = \underline{m}\;\right] = \left(L'\hat{\underline{\beta}}_\Gamma - \underline{m}\;\right)'\left(L'(X'X)_\Gamma^{+}L\right)^{-1}\left(L'\hat{\underline{\beta}}_\Gamma - \underline{m}\;\right), \tag{5.58}$$

   where $\hat{\underline{\beta}}_\Gamma$ is any LSS to the normal equations.

2. For $\breve{L}'\underline{\beta} = \underline{m}$, a testable hypothesis in the model constrained by $K'\underline{\beta} = \underline{0}$ where Col(K) and Col($\breve{L}$) are linearly independent,

$$SS\left[H_0\!: \breve{L}'\underline{\beta} = \underline{m} \mid K'\underline{\beta} = \underline{0}\right] = \left(\breve{L}'\hat{\underline{\beta}}_\Gamma - \underline{m}\;\right)'\left(\breve{L}'(X'X)_\Gamma^{+}\breve{L}\right)^{-1}\left(\breve{L}'\hat{\underline{\beta}}_\Gamma - \underline{m}\;\right), \tag{5.59}$$

   where $\hat{\underline{\beta}}_\Gamma$ is any LSS satisfying $K'\underline{\beta} = \underline{0}$.

3. For $\tilde{L}'\tilde{\underline{\beta}} = \tilde{\underline{m}}$, a testable hypothesis in the reparameterized reduced model (5.10),

$$SS\left[H_0\!: \tilde{L}'\tilde{\underline{\beta}} = \tilde{\underline{m}}\;\right] = \left(\tilde{L}'\hat{\tilde{\underline{\beta}}} - \tilde{\underline{m}}\;\right)'\left(\tilde{L}'(\tilde{K}'X'X\tilde{K})_\Gamma^{+}\tilde{L}\right)^{-1}\left(\tilde{L}'\hat{\tilde{\underline{\beta}}} - \tilde{\underline{m}}\;\right), \tag{5.60}$$

   where $\hat{\tilde{\underline{\beta}}}_\Gamma$ is any LSS to the normal equations corresponding to the {1,2}-inverse $(\tilde{K}'X'X\tilde{K})_\Gamma^{+}$.

4. For contrasts in terms of the cell sample means,

$$SS[H_0: C'\underline{\mu} = \underline{m}\ ] = \left(C'\underline{z} - \underline{m}\right)'\left(C'D^{-1}C\right)^{-1}\left(C'\underline{z} - \underline{m}\right), \tag{5.61}$$

where $\underline{z}$ is the vector of cell sample means and D is the diagonal matrix of cell sizes.

5. For contrasts in terms of the cell means model,

$$SS[H_0: M'\underline{\mu} = \underline{m} \mid G'\underline{\mu} = \underline{0}] = \left(M'\hat{\underline{\mu}}_{\Phi} - \underline{m}\right)'\left(M'D_{\Phi}^{+}M\right)^{-1}\left(M'\hat{\underline{\mu}}_{\Phi} - \underline{m}\right), \tag{5.62}$$

where $\Phi = \text{Col}(X_*)$ and X_* is the cell model matrix.

•

5.4 Equivalent Numerator Sum of Squares For a Test of Hypothesis

In the previous section, five different approaches for computing numerator sums of squares were given in Corollary 5.5.1. Each depended on a different representation of the underlying model or a different analytical approach. In this section, sufficient conditions on L', $\breve{L}'$, $\tilde{L}'$, M', and C' will be given which assure the equality of the numerator sums of squares (5.58) through (5.62).

The choice of such contrasts is not a trivial matter. There are some subtle considerations which complicate the search for equivalent hypotheses.

As pointed out by Hackney (1976), when imposing a set of nonestimable constraints $K'\underline{\beta} = \underline{0}$ where $\text{Rank}(K) > p-r$, one is in fact imposing some equivalent estimable constraints. For example: in the 2^2 factorial model, imposing the nonestimable constraints

$$\gamma_{ij} = 0, \text{ for all } i, j \tag{5.63}$$

on the interactions is equivalent to imposing the estimable constraints

$$\gamma_{ij} - \gamma_{i'j} - \gamma_{ij'} + \gamma_{i'j'} = 0. \tag{5.64}$$

She went on to show that in such a constrained model, the hypothesis H_0: $L'\underline{\beta} = \underline{0}$ is equivalent, in the unconstrained model, to the hypothesis

$$H_0: (K' + HL')\underline{\beta} = \underline{0} \tag{5.65}$$

for an appropriate H. In her thesis, she gave choices for H where the Σ-constraints were imposed in the two-way model.

While hypotheses in the Constrained and Cell Means Models must be consistent with the constraints of those models, hypotheses in terms of the cell sample means are not so restricted. Any hypothesis H_0: $C'\underline{\mu} = \underline{m}$ is testable with the appropriate numerator sum of squares given by (5.61). Thus, the set of hypotheses tested in terms of the unconstrained cell sample means is larger than that testable in terms of some unsaturated parameterized linear model. As pointed out by Hsuan (1982), for (5.61) to be equivalent to a hypothesis tested in the unconstrained model, it is necessary and sufficient that

$$G'D^{-1}C = 0 \tag{5.66}$$

or equivalently, by (5.5), that

$$\text{Col}(C) \subset \text{Col}(W'X). \tag{5.67}$$

The ultimate aim of this section is to develop a series of algorithms for equating the hypotheses tested by the various analytical approaches. The strategy used here will be that suggested by Hsuan (1982).

The following theorem establishes sufficient conditions when the various numerator sum of squares of Corollary 5.5.1 are equal.

Theorem 5.6: For the model $\underline{y} = X\underline{\beta}+\underline{\epsilon}$ constrained by $K'\underline{\beta} = \underline{0}$ with $\hat{\underline{\beta}}_\Gamma$ any LSS satisfying $K'\hat{\underline{\beta}}_\Gamma = \underline{0}$ let W, X_*, $\tilde{K}$, D, G and Φ be as defined in Section 5.1.

1. For any Q′ the following are equal:

$$SS\left[H_0\text{: } Q'(X'X)_\Gamma^+X'X\underline{\beta} = \underline{0}\right], \tag{5.68}$$

$$SS\left[H_0\text{: } Q'(X'X)_\Gamma^+X'X\underline{\beta} = \underline{0} \mid K'\underline{\beta} = \underline{0}\right], \tag{5.69}$$

$$SS\left[H_0\text{: } Q'(X'X)_\Gamma^+X'X\tilde{K}\tilde{\underline{\beta}} = \underline{0}\right], \tag{5.70}$$

$$SS\left[H_0\text{: } Q'(X'X)_\Gamma^+X'W\underline{\mu} = \underline{0}\right], \text{ and} \tag{5.71}$$

$$SS\left[H_0\text{: } Q'(X'X)_\Gamma^+X'W\underline{\mu} = \underline{0} \mid G'\underline{\mu} = \underline{0}\right]. \tag{5.72}$$

2. For $\tilde{L}'\underline{\beta}$ estimable the following are equal:

$$SS\left[H_0\text{: } \tilde{L}'\underline{\beta} = \underline{0} \mid K'\underline{\beta} = \underline{0}\right], \tag{5.73}$$

$$SS\left[H_0\text{: } \tilde{L}'(X'X)_\Gamma^+X'X\underline{\beta} = \underline{0} \mid K'\underline{\beta} = \underline{0}\right], \tag{5.74}$$

$$SS\left[H_0\text{: } \tilde{L}'(X'X)_\Gamma^+X'X\underline{\beta} = \underline{0}\right], \tag{5.75}$$

$$SS\left[H_0\colon\ \tilde{L}'(X'X)^+_{\Gamma}X'X\tilde{K}'\tilde{\underline{\beta}} = \underline{0}\right], \tag{5.76}$$

$$SS\left[H_0\colon\ \tilde{L}'(X'X)^+_{\Gamma}X'W\underline{\mu} = \underline{0} \mid G'\underline{\mu} = \underline{0}\right], \text{ and} \tag{5.77}$$

$$SS\left[H_0\colon\ \tilde{L}'(X'X)^+_{\Gamma}X'W\underline{\beta} = \underline{0}\right]. \tag{5.78}$$

3. For any C′ the following are equal:

$$SS\left[H_0\colon\ C\underline{\mu} = \underline{0} \mid G'\underline{\mu} = \underline{0}\right], \tag{5.79}$$

$$SS\left[H_0\colon\ C'D^+_{\Phi}D\underline{\mu} = \underline{0} \mid G'\underline{\mu} = \underline{0}\right], \text{ and} \tag{5.80}$$

$$SS\left[H_0\colon\ C'D^+_{\Phi}D\underline{\mu} = \underline{0}\right]. \tag{5.81}$$

Proof: From Lemmas 5.1, 5.2 and 5.3 the following equalities hold:

$$\hat{\underline{\beta}}_{\Gamma} = (X'X)^+_{\Gamma}X'X\hat{\underline{\beta}}_{\Gamma} \tag{5.82}$$

$$= (X'X)^+_{\Gamma}X'X\hat{\underline{\beta}} \tag{5.83}$$

$$= \tilde{K}'\hat{\tilde{\underline{\beta}}}_{\Psi} \tag{5.84}$$

$$= (X'X)^+_{\Gamma}X'X\tilde{K}\hat{\tilde{\underline{\beta}}}_{\Psi} \tag{5.85}$$

$$= (X'X)^+_{\Gamma}X'W\hat{\underline{\mu}}_{\Phi} \tag{5.86}$$

$$= (X'X)^+_{\Gamma}X'W\underline{z} \tag{5.87}$$

where $\Psi = \left(\tilde{K}'\tilde{K}\right)^{-1}\tilde{K}'(\Gamma)$. Col(K) and $\mathrm{Col}\left(X'X(X'X)^+_{\Gamma}\right)$ are linearly independent. Supposing otherwise, there must exists matrices P and Q such that

$$PK' = Q(X'X)^+_{\Gamma}X'X. \tag{5.88}$$

Thus

$$\{\underline{0}\} = Q(\Gamma) = Q(X'X)^+_{\Gamma}X'X(\Gamma) \tag{5.89}$$

which implies

$$0 = Q(X'X)^+_{\Gamma}X'X. \tag{5.90}$$

Likewise Col(G) and $\mathrm{Col}\left(W'X(X'X)^+_{\Gamma}\right)$ are linearly independent, as are Col(G) and $\mathrm{Col}(DD^+_{\Psi})$. Part 1 follows immediately by multiplying the equations (5.82) through (5.87) on the left by Q′ and then applying Corollary 5.5.1. Part 2 follows similarly by multiplying the same equations on the left by $\tilde{L}'$ and again applying the same corollary. From Lemma 5.2 the following equality holds:

$$\underline{\mu}_{\Phi} = D^+_{\Phi}D\underline{\mu}_{\Phi} = D^+_{\Phi}D\underline{z}. \tag{5.91}$$

Part 3 follows from corollary 5.5.1 by premultiplying (5.91) by C′.

•

The needed quantities in the theorem are easily obtained. From Lemma 2.10 it may be easily verified that

$$(X'X)^{+}_{\Gamma} = \tilde{K}(\tilde{K}'X'X\tilde{K})^{-}\tilde{K}' \qquad (5.92)$$

and

$$D^{+}_{\Phi}D = X_{*}(X'X)^{-}X'W \qquad (5.93)$$

for arbitrary g-inverses.

Theorem 5.6 provides the basis for a series of algorithms for determining the equivalent hypotheses tested by the different analytical approaches. The algorithms vary depending upon one's starting point.

<u>Algorithm 1</u>: Part 1 of the Theorem may be represented symbolically as

$$(X'X)^{+}_{\Gamma}X'\left[X \mid W \mid X\tilde{K}\right] \xrightarrow[\text{operations}]{\text{row}} \left[L' \mid C' \mid \tilde{L}'\right] \qquad (5.94)$$

which implies

$$\begin{aligned} SS\left[H_0\colon L'\underline{\beta} = \underline{0}\right] &= SS\left[H_0\colon L'\underline{\beta} = \underline{0} \mid K'\underline{\beta} = \underline{0}\right] \\ &= SS\left[H_0\colon \tilde{L}'\underline{\tilde{\beta}} = \underline{0}\right] \\ &= SS\left[\ H_0\colon C'\underline{\mu} = \underline{0}\right] \\ &= SS\left[\ H_0\colon C'\underline{\mu} = \underline{0} \mid G'\underline{\mu} = \underline{0}\right]. \end{aligned} \qquad (5.95)$$

•

If one's starting point is a constrained model, the hypothesis, $\tilde{L}'\underline{\beta} = \underline{0}$, is often given. It is desirable to avoid looking for some Q′ such that $Q'(X'X)^{+}_{\Gamma}X'X = \tilde{L}'$. Such a matrix Q′ may, in fact, not exist. As noted in the introduction to Theorem 5.6, an arbitrary testable hypothesis, although nominally testable, is in reality, due to the constraints on the model, equivalent to some other hypothesis which is linearly independent of the constraints. This alternate hypothesis together with those in the other four approaches may be found as follows:

<u>Algorithm 2</u>: Given the hypothesis in the constrained model $\tilde{L}'\underline{\tilde{\beta}} = \underline{0}$, let

$$\tilde{L}'(X'X)^{+}_{\Gamma}X'\left[X \mid W \mid \tilde{K}\ \right] = \left[L' \mid C' \mid \tilde{L}'\right], \qquad (5.96)$$

then

$$\begin{aligned} SS\left[H_0\!: \tilde{L}'\underline{\beta} = \underline{0} \mid K'\underline{\beta} = \underline{0}\right] &= SS\left[H_0\!: L'\underline{\beta} = \underline{0} \mid K'\underline{\beta} = \underline{0}\right] \\ &= SS\left[H_0\!: L'\underline{\beta} = \underline{0}\right] \\ &= SS\left[H_0\!: \tilde{L}'\underline{\tilde{\beta}} = \underline{0}\right] \qquad (5.97) \\ &= SS\left[\ H_0\!: C'\underline{\mu} = \underline{0}\right] \\ &= SS\left[\ H_0\!: C'\underline{\mu} = \underline{0} \mid G'\underline{\mu} = \underline{0}\right]. \end{aligned}$$

●

In the reparameterized reduced model, a linear function of the parameters $\tilde{L}'\underline{\tilde{\beta}} = \underline{0}$ is estimable if and only if $\tilde{L}' = Q'X\tilde{K}$ for some Q'. In particular, Q' can be chosen such that

$$\mathrm{Col}(Q) \subset \mathrm{Col}(X\tilde{K}) = X(\Gamma). \qquad (5.98)$$

If (5.98) holds, then since $X(X'X)_{\Gamma}^{+}X'$ is a projection to $X(\Gamma)$, it follows that

$$Q' = Q'X(X'X)_{\Gamma}^{+}X'. \qquad (5.99)$$

Thus the following corollary follows trivially from part 1 of Theorem 5.6:

Corollary 5.6.1: When $\mathrm{Col}(Q) \subset \mathrm{Col}(X\tilde{K})$ then

$$\begin{aligned} &SS\left[H_0\!: Q'X\tilde{K}'\underline{\tilde{\beta}} = \underline{0}\right] \\ &\quad = SS\left[H_0\!: Q'X\underline{\beta} = \underline{0}\right] \\ &\quad = SS\left[H_0\!: Q'X\underline{\beta} = \underline{0} \mid K'\underline{\beta} = \underline{0}\right] \qquad (5.100) \\ &\quad = SS\left[\ H_0\!: Q'W\underline{\mu} = \underline{0}\right] \\ &\quad = SS\left[\ H_0\!: Q'W\underline{\mu} = \underline{0} \mid G'\underline{\mu} = \underline{0}\right]. \end{aligned}$$

●

The above corollary may be restated in the following way. Given a testable hypothesis in the reparameterized reduced model, the corresponding hypothesis in terms of the unconstrianed model, the constrained model, the cell means model, or the cell sample means is expressed by taking the "same" linear combinations of the rows of the corresponding design matrices X, W, and $X\tilde{K}$.

Algorithm 3: Let $\tilde{L}'\underline{\beta} = \underline{0}$ be a testable hypothesis in the reparameterized reduced model. Write the contrasts as $\tilde{L}' = Q'X\tilde{K}$, where $\mathrm{Col}(Q) \subset \mathrm{Col}(X\tilde{K})$ then

$$Q'\left[X \mid W \mid X\tilde{K}\right] = \left[L' \mid C' \mid \tilde{L}'\right] \qquad (5.101)$$

which implies

$$
\begin{aligned}
SS\left[H_0\colon \tilde{L}'\tilde{\underline{\beta}} = \underline{0}\right] &= SS\left[H_0\colon L'\underline{\beta} = \underline{0}\right] \\
&= SS\left[H_0\colon \tilde{L}'\underline{\beta} = \underline{0} \mid K'\underline{\beta} = \underline{0}\right] \\
&= SS\left[\, H_0\colon C'\underline{\mu} = \underline{0}\right] \\
&= SS\left[\, H_0\colon C'\underline{\mu} = \underline{0} \mid G'\underline{\mu} = \underline{0}\right].
\end{aligned} \tag{5.102}
$$

●

One application of Algorithm #3 is in the reduction of squares which will be explored in Section 5.6.

As a final application of Theorem 5.5, attention will be focused specifically on the equivalence of hypotheses in the unconstrained model, the cell means model and in terms of cell sample means. In this case, $\mathrm{Dim}(\Gamma) = \mathrm{Rank}(X)$. It follows that for any estimable function

$$L'\underline{\beta} = L'(X'X)^{+}_{\Gamma}X'X\underline{\beta}. \tag{5.103}$$

Therefore the equivalent contrast in terms of the cell means is

$$C'\underline{\mu} = L'(X'X)^{+}_{\Gamma}X'W\underline{\mu} \tag{5.104}$$

regardless of the requirement that $G'\underline{\mu} = \underline{0}$. Conversely, for any matrix C with $\mathrm{Col}(C) \subset \mathrm{Col}(W'X)$, there exists a matrix Q such that

$$
\begin{aligned}
C' &= Q'X'W = Q'X'WX_*(X'X)^{+}_{\Gamma}X'W \\
&= C'X_*(X'X)^{+}_{\Gamma}X'W \\
&= C'X_*(X_*'DX_*)^{+}_{\Gamma}X_*'D \\
&= C'D^{+}_{\Phi}D.
\end{aligned} \tag{5.105}
$$

From (5.68) and (5.72) of Theorem 5.6, for the contrast $C'\underline{\mu} = \underline{0}$ in the cell means model, the equivalent contrast in the unconstrained model is

$$L'\underline{\beta} = C'X_*(X'X)^{+}_{\Gamma}X'X\underline{\beta} = C'D^{+}_{\Phi}DX_*\underline{\beta}. \tag{5.106}$$

Expressions (5.104) and (5.106) together form the basis for the following:

<u>Algorithm 4</u>: Let $K'\underline{\beta} = \underline{0}$ be a set of $p - r$ linearly independent non-estimable constraints and Γ be the orthogonal complement of Col(K), then for any estimable functions $L'\underline{\beta}$ and $C'\underline{\mu}$ where $\mathrm{Col}(C) \subset \mathrm{Col}(W'X)$, either

$$C' = L'(X'X)^{+}_{\Gamma}X'W \tag{5.107}$$

or

$$L' = C'X_*(X'X)^{+}_{\Gamma}X'X \tag{5.108}$$

implies $$SS[H_0\colon L'\underline{\beta} = \underline{0}] = SS[H_0\colon L'\tilde{K}\underline{\tilde{\beta}} = \underline{0}]$$
$$= SS[\ H_0\colon C'\underline{\mu} = \underline{0}] \qquad (5.109)$$
$$= SS[\ H_0\colon C'\underline{\mu} = \underline{0} \mid G'\underline{\mu} = \underline{0}].$$

•

The expressions (5.107) and (5.108) apparently differ from those found in Schmoyer (1984) which in terms of the present notation are

$$C'\underline{\mu} = L'(X'X)^{-}X'W\wp_{\Phi}\underline{\mu}\,, \qquad (5.110)$$

and

$$L'\underline{\beta} = C'\wp_{\Phi}\left(\wp'_{\Phi}D\wp_{\Phi}\right)^{-}\wp'_{\Phi}W'X\underline{\beta}\,, \qquad (5.111)$$

where $\wp_{\Phi}$ is the Euclidean projection into $\Phi = \mathrm{Col}(X_*)$ and the g-inverses are arbitrary. However, it is assumed that $G'\underline{\mu} = \underline{0}$ and $\Phi = \mathrm{Col}(G)^{\perp}$. Therefore it follows that $\wp_{\Phi}\underline{\mu} = \underline{\mu}$. Thus (5.104) and (5.110) are equivalent. Since by Lemma 2.10,

$$D_{\Phi}^{+} = \wp_{\Phi}\left(\wp'_{\Phi}D\wp_{\Phi}\right)^{-}\wp'_{\Phi} \qquad (5.112)$$

it follows that (5.106) and (5.111) are equal.

With this in mind, the results of this section may be thought of as a generalization of the results of Schmoyer to the case of the constrained linear model.

5.5 Hypotheses Invariant to Cell Sizes

As mentioned in the Introduction, as long as there are no empty cells, when the "usual" restrictions are imposed and the R^* approach is used to generate a hypothesis for the various effects, the hypotheses expressed in terms of the parameters of the original overparameterized model do not depend upon the cell sizes. For example, in the two way interaction the hypothesis for the α-effect tested by the R^*-approach is

$$H_0\colon\ \alpha_i + \sum_j \gamma_{ij} = \alpha_{i'} + \sum_j \gamma_{i'j}\,,\ \text{for all } i,\ i'. \qquad (5.113)$$

This same hypothesis expressed in terms of the cell means is also invariant to the cell sizes:

$$H_0\colon\ \underline{\mu}_{i+} = \underline{\mu}_{i'+}\ \text{for all } i,\ i'. \qquad (5.114)$$

It is apparent from the previous section that the key expression for the respresentation of the tested hypothesis in terms of the parameters of the original model, constrained or otherwise, is

$$(X'X)^+_\Gamma X'X. \tag{5.115}$$

Similarly

$$(X'X)^+_\Gamma X'W \tag{5.116}$$

is the key expression in the representation of the tested hypothesis in terms of the cell means. The invariance of each of these expressions to the cell sample sizes will lead to the invariance of the hypothesis as expressed in terms of either the parameters unconstrained model or in terms of the cell means. The following lemma summarizes when (5.115) and (5.116) will be invariant to the cell sample sizes.

Lemma 5.7: Assuming no empty cells.

1. If $\mathrm{Dim}(\Gamma) = r$ then
$$(X'X)^+_\Gamma X'X = (X_*'X_*)^+_\Gamma X_*'X_* \, . \tag{5.117}$$
2. If $\mathrm{Dim}(\Gamma) = r$ and the design is saturated then
$$(X'X)^+_\Gamma X'W = (X_*'X_*)^+_\Gamma X_*'W \, . \tag{5.118}$$

Proof:

1. $(X'X)^+_\Gamma X'X$ is the projection to Γ along $[X'X(\Gamma)]^\perp$. If $\mathrm{Dim}(\Gamma) = r$ then $\mathrm{Row}(X_*) = X'X(\Gamma)$.
2. $(X'X)^+_\Gamma X'W$ is the {2}-inverse of X_* with column space Γ and row space $DX_*(\Gamma)$. Under the conditions of the lemma, $\Re^n = DX_*(\Gamma.)$

●

The above lemma outlines sufficient conditions for hypotheses to be invariant to the cell sizes. The proof of the necessity of the conditions is not apparent to this author. In the example at the end of this section, it will be demonstrated that, for the two-way model without interaction, if one uses the "usual" constraints, the hypotheses tested in the resulting full rank model, when expressed in terms of original overparameterized model, do not depend upon the cell sizes. In contrast, the same hypotheses when expressed in terms of the cell means do vary with the cell sizes. This point has not been made clear in the literature, since the model most commonly used for illustration is the saturated two-way interaction model.

Example 5.1: Recall for the unsaturated model (4.32) and the data (4.33) of Example 4.2 in the previous chapter, corresponding to the "usual" constraints $\alpha_1+\alpha_2 = 0$ and $\beta_1+\beta_2 = 0$, we had

$$\hat{\underline{\beta}}_{\Gamma_1} = (X'X)^{+}_{\Gamma_1} X'\underline{y} = \frac{1}{2}\begin{bmatrix} 33 \\ -2 \\ 2 \\ 5 \\ -5 \end{bmatrix}, \tag{5.119}$$

where

$$(X'X)^{+}_{\Gamma_1} = \frac{1}{28}\begin{bmatrix} 6 & -1 & 1 & -1 & 1 \\ -1 & 6 & -6 & -1 & 1 \\ 1 & -6 & 6 & 1 & -1 \\ -1 & -1 & 1 & 6 & -6 \\ 1 & 1 & -1 & -6 & 6 \end{bmatrix}. \tag{5.120}$$

Thus

$$(X'X)^{+}_{\Gamma_1} X'W = \frac{1}{28}\begin{bmatrix} 8 & 6 & 6 & 8 \\ 8 & 6 & -8 & -6 \\ -8 & -6 & 8 & 6 \\ 8 & -8 & 6 & -6 \\ -8 & 8 & -6 & 6 \end{bmatrix} \tag{5.121}$$

and

$$(X'X)^{+}_{\Gamma_1} X'X = \frac{1}{2}\begin{bmatrix} 2 & 1 & 1 & 1 & 1 \\ 0 & 1 & -1 & 0 & 0 \\ 0 & -1 & 1 & 0 & 0 \\ 0 & 0 & 0 & 1 & -1 \\ 0 & 0 & 0 & -1 & 1 \end{bmatrix}. \tag{5.122}$$

In contrast, with only one observation per cell

$$X'X = \begin{bmatrix} 4 & 2 & 2 & 2 & 2 \\ 2 & 2 & 0 & 1 & 1 \\ 2 & 0 & 2 & 1 & 1 \\ 2 & 1 & 1 & 2 & 0 \\ 2 & 1 & 1 & 0 & 2 \end{bmatrix} \tag{5.123}$$

and

$$(X'X)^{+}_{\Gamma_1} = \frac{1}{4}\begin{bmatrix} 1 & 0 & 0 & 0 & 0 \\ 0 & 1 & -1 & 0 & 0 \\ 0 & -1 & 1 & 0 & 0 \\ 0 & 0 & 0 & 1 & -1 \\ 0 & 0 & 0 & -1 & 1 \end{bmatrix}. \tag{5.124}$$

For this model

$$(X'X)^{+}_{\Gamma_1}X'W = \frac{1}{4}\begin{bmatrix} 1 & 1 & 1 & 1 \\ 1 & 1 & -1 & -1 \\ -1 & -1 & 1 & 1 \\ 1 & -1 & 1 & -1 \\ -1 & 1 & -1 & 1 \end{bmatrix}, \tag{5.125}$$

which does not agree with (5.121), and

$$(X'X)^{+}_{\Gamma_1}X'X = \frac{1}{2}\begin{bmatrix} 2 & 1 & 1 & 1 & 1 \\ 0 & 1 & -1 & 0 & 0 \\ 0 & -1 & 1 & 0 & 0 \\ 0 & 0 & 0 & 1 & -1 \\ 0 & 0 & 0 & -1 & 1 \end{bmatrix}, \tag{5.126}$$

which agrees with (5.122).

•

5.6 The R approach and SAS Type I and Type II Sums of Squares

In the R appoach the interest is in computing the sum of squares for an effect, say $\underline{\beta}_2$, after "adjusting" for the effects $\underline{\beta}_1$, but ignoring other effects, say $\underline{\beta}_3$.

Let the design matrix be partitioned as

$$X = \left[\, X_1 \mid X_2 \mid X_3 \,\right] \tag{5.127}$$

and the parameters conformably as

$$\underline{\beta}' = \left[\, \underline{\beta}_1' \mid \underline{\beta}_2' \mid \underline{\beta}_3' \,\right]. \tag{5.128}$$

If $\text{Res}(\underline{\beta}_1,\underline{\beta}_2)$ is the residual sum of squares for the model

$$\underline{y} = [X_1 \mid X_2]\begin{bmatrix} \underline{\beta}_1 \\ \underline{\beta}_2 \end{bmatrix} + \underline{\epsilon}, \tag{5.129}$$

and $\text{Res}(\underline{\beta}_1)$ is the residual sum of squares for the model

$$\underline{y} = X_1\underline{\beta}_1+\underline{\epsilon}\ , \tag{5.130}$$

then the sum of squares due to fitting $\underline{\beta}_2$ after $\underline{\beta}_1$ and ignoring $\underline{\beta}_3$ is given by

$$R(\underline{\beta}_2|\underline{\beta}_1) = Res(\underline{\beta}_1) - Res(\underline{\beta}_1,\underline{\beta}_2). \tag{5.131}$$

As indicated in the SAS User's Manual (1985), the hypothesis tested by (5.132) corresponds to the Type I and Type II sum of squares for $\underline{\beta}_2$ when $\underline{\beta}_1$ appears first in the model statement followed by $\underline{\beta}_2$ and then $\underline{\beta}_3$. It is given by $L'\underline{\beta} = \underline{0}$ for

$$L' = X_2' M_1 X \ \left[0 \mid X_2' M_1 X_2 \mid X_2' M_1 X_3\right] \tag{5.132}$$

where $M_1 = I - X_1(X_1' X_1)^{-} X_1'$ for any g-inverse. One approach in verifying this begins by noting that the hypothesis tested by (5.131) expressed in terms of the reduced model (5.129) is given by

$$\underline{0} = \tilde{L}'\begin{bmatrix}\underline{\beta}_1\\ \underline{\beta}_2\end{bmatrix} \tag{5.133}$$

where $\tilde{L}'$ is any estimable function such that

$$Row(\tilde{L}') = \left\{\left[\ \underline{0}' \mid \tilde{\underline{\lambda}}_2'\ \right] \in Row\left(\left[X_1 \mid X_2\right]\right) \,\middle|\, \tilde{\underline{\lambda}}_2' \neq \underline{0}'\right\}. \tag{5.134}$$

One choice for L' is

$$\tilde{L}' = X_2' M_1\left[X_1 \mid X_2\right] = \left[0 \mid X_2' M_1 X_2\right]. \tag{5.135}$$

Since $Col\left(\left[M_1X_2\right]\right) \subset Col\left(\left[\ X_1 \mid X_2\right]\right)$ it follows from Algorithm #3 that (5.132) is the linear function defining the hypothesis tested by (5.131).

5.7 The R* Approach And The SAS Type III Sum of Squares

The Σ-, W- and 0-constraints discussed in section 5.1 share a particular form. In each case, the restrictions involve only the parameter of each effect separately. Thus the K associated with the reparameterization is a block diagonal matrix with one block for each effect. More precisely, suppose in model (5.1) there are k effects, then let

$\underline{\beta}' = \left[\ \underline{\beta}_1',\ \underline{\beta}_2',\cdots,\underline{\beta}_k'\right]$ where $\underline{\beta}_i$ is the p_ix1 vector of parameters associated with the i-th effect,

$\tilde{K}$ = Diag$[\tilde{K}_1, \tilde{K}_2, \cdots, \tilde{K}_k]$ where $\tilde{K}_t$, corresponding to $\tilde{\underline{\beta}}_t$, is of order $p_t x r_t$ and of rank r_t,

$\tilde{X}$ = $X\tilde{K}$ = $[X_1\tilde{K}_1, X_2\tilde{K}_2, \cdots, X_k\tilde{K}_k]$ be full column rank $r = \sum_t r_t$,

$\tilde{\underline{\beta}}_t$ = the $r_t x1$ vector of parameters associated with X_tK_t, and

$\tilde{\underline{\beta}}' = [\tilde{\underline{\beta}}'_1, \tilde{\underline{\beta}}'_2, \cdots, \tilde{\underline{\beta}}'_k]$.

In the resulting full rank reparameterized model, $\tilde{\underline{\beta}}_t$ is estimable. Since $\tilde{K}_t$ is full column rank

$$\tilde{\underline{\beta}}_t = \underline{0} \text{ if and only if } \tilde{K}_t\tilde{\underline{\beta}}_t = \underline{0}. \tag{5.136}$$

Let Γ = Col($\tilde{K}$), then

$$\tilde{K} = (X'X)^+_\Gamma X'X\tilde{K}\ . \tag{5.137}$$

From part 1 of Theorem 5.6 and the above two expressions it follows that

$$SS\left[Ho\colon \tilde{\underline{\beta}}_t = \underline{0}\right] = SS\left[H_0\colon \left[(X'X)^+_\Gamma X'X\right]_t \underline{\beta} = \underline{0}\right], \tag{5.138}$$

where $\left[(X'X)^+_\Gamma X'X\right]_t$ indicates the rows of $(X'X)^+_\Gamma X'X$ associated with $\underline{\beta}_t$.

If Γ represents the space orthogonal to that generated by either the Σ-, W- or 0-constraints, then by Lemma 5.7

$$(X'X)^+_\Gamma X'X = (X'_* X_*)^+_\Gamma X'_* X_*. \tag{5.139}$$

Consequently, (5.139) represents an efficient way to express in terms of the original overparameterized model the hypothesis tested by the R^* approach whenever the Σ-, W- or 0-constraints are used. In contrast, one can look at the description of the SAS Type III sums of squares found in the SAS User's Guide Statistics [(1985), p. 88].

REFERENCES

Aitken, A. C.: On The Statistical Independence of Quadratic Forms In Normal Variates., Biometrika 37, 93-96 (1950)

Baldesari, B.: The Distribution of A Quadratic Form Of Normal Random Variables., Ann. Stat. 38, 1700-1704 (1967)

Ben-Israel, A.; Greville, T. N. E.: Generalized Inverses: Theory and Applications, New York, Wiley (1974)

Bhat, H. W.: On The Distribution of Certain Quadratic Forms in Normal Variates., J. R. Stat. Soc., Ser. B 24, 148-151 (1962)

Bott, R.; Duffin, R. J.: On The Algebra of Networks., Trans. Am. Math. Soc. 74, 99-109 (1953)

Carpenter, O.: Note On The Extension of Craig's Theorem to Non-Central Variates., Ann. Stat. 21, 455-457 (1950)

Cochran, W. G.: The Distribution of Quadratic Forms in a Normal System with Applications to the Analysis of Variance., Proc. Camb. Phil. Soc. 30, 178-191 (1934)

Craig, A. T.: Note On the Independence of Certain Quadratic Forms., Ann. Stat. 14, 195-197 (1943)

Craig, A. T.: Bilinear Forms in Normal Correlated Variables., Ann. Stat. 18, 565-573 (1947)

Draper, N. R.; Smith, H.: Applied Regression Analysis, New York, Wiley (1981)

Fisher, R. A.: Applications of 'Student's' Distribution., Metron 5, 90-104 (1926)

Freund, R. J.; Littell, R. C.: SAS For Linear Models: A Guide to the ANOVA and GLM Precedures., Cary, SAS Institute Inc. (1981)

Ghurye, S. G.; Olkin, I.: A Characterization Of The Multivariate Normal Distribution., Ann. Stat. 33, 533-541 (1962)

Golub, G. H.; Van Loan, C. F.: Matrix Computations., Baltimore, John Hopkins University Press (1983)

Good, I. J.: On The Independence of Quadratic Expressions., J. R. Stat. Soc., Ser. B 25, 377-382 (1963)

Good, I. J.: Conditions For A Quadratic Form to Have Chi-Square Distribution., Biometrika 56, 215-216 (1969)

Good, I. J.: Conditions For A Quadratic Form to Have Chi-Square Distribution. Correction., Biometrika 57, 225 (1970)

Goodnight, J. H.: Tests of Hypothesis in Fixed Effects Linear Models., SAS Technical Report R-101, Cary, SAS Institute Inc. (1978)

Goodnight, J. H.: A Tutorial on the SWEEP Operator., Am. Stat. 33, 149-158 (1979)

Graybill, F. A.: Theory and Application of the Linear Model, Belmont, Wadsworth Publishing Co. (1976)

Graybill, F. A.; Marsaglia, G.: Idempotent Matrices And Quadratic Forms in The General Linear Hypothesis., Ann. Stat. 28, 678-686 (1957)

Grizzle, J. E.; Starmer, C. F.; Koch, G. G.: Analysis of Categorical Data by Linear Models., Biometrics 25, 489-504 (1969)

Hackney, O. P.: Hypothesis Testing In the General Linear Model., Ph.D. Dissertation, Mississippi State University (1976)

Hocking, R. R.; Speed, F. M.: A Full Rank Analysis of Some Linear Model Problems., J. Am. Stat. Assoc. 70, 706-712 (1975)

Hogg, R. V.; Craig, A. T.: On The Decomposition of Certain Chi-Squared Variables., Ann. Stat. 29, 608-610 (1958)

Hotelling, H.: Note On A Matrix Theorem of A. T. Craig., Ann. Stat. 15, 427-429 (1944)

Hsuan, F. C.: The Equivalence of GLM and Cell Means Models In The Construction of Analysis of Variance Tables., 142nd Annual Joint Statistical Meetings, August 16-19, Cincinnati, Ohio (1982)

Hsuan, F.; Langenberg, P.; Getson, A.: The {2}-Inverse With Applications in Statistics., Linear Algebra Appl. 70, 241-248 (1985)

Kac, M.: A Remark On The Independence of Linear And Quadratic Forms Involving Gaussian Variables., Ann. Stat. 16, 400-401 (1945)

Khatri, C. G.: Conditions For Wishartness And Independence Of Second Degree Polynomials In A Normal Vector., Ann. Stat. 33, 1002-1007 (1962)

Khatri, C. G.: Further Contributions to Wishartness And Independence Of Second Degree Polnomials In Normal Variates., J. Indian Stat. Assoc. 1, 61-70 (1963)

Khatri, C. G.: Quadratic Forms And Extension of Cochran's Theorem To Normal Vector Variables., In P.R. Krishnaiah ed., Multivariate Analysis, Vol. IV, North-Holland Publishing Company, 79-94 (1977)

Khatri, C. G.: A Remark On The Necessary and Sufficient Conditions For A Quadratic Form To Be Distributed As Chi-Squared., Biometrika 65, 239-240 (1978)

Khatri, C. G.: Quadratic Forms In Normal Variables., In P.R. Krishnaiah ed., Handbook Of Statistics, Vol. I, North-Holland Publishing Company, 443-469 (1980)

Kruskal, W.: The Geometry of Generalized Inverses., J. R. Stat. Soc., Ser. B 37, 272-283 (1975)

Kutner, M. H.: Hypothesis Testing in Linear Models (Eisenhart Model I)., Am. Stat. 28, 98-100 (1974)

Laha, R. G.: On The Stochastic Independence Of Two Second-Degree Polynomial Statistics In Normally Distributed Variates., Ann. Stat. 27, 790-796 (1956)

Luther, N. Y.: Decomposition Of Symmetric Matrices And Distributions Of Quadratic Forms., Ann. Stat. 36, 683-690 (1965)

Madow, W. G.: The Distribution of Quadratic Forms In Non-Central Normal Random Variables., Ann. Stat. 11, 101-103 (1940)

Matern, B.: Independence of Non-Negative Quadratic Forms In Normally Correlated Variables., Ann. Stat. 20, 119-120 (1949)

Mazumdar, S.; Li, C. C.; Bryce, G. R.: Correspondence Between a Linear Restriction and a Generalized Inverse in Linear Model Analysis., Am. Stat. 34, 103-105 (1980)

Mitra, S. K.: On A Generalized Inverse Of A Matrix And Applications., Sānkhyā, Ser. A 30, 107-114 (1968)

Monlezun, C. J.; Speed, F. M.: The Geometry of Estimation and Hypothesis Testing in the Constrained Linear Model - The Full Rank Case., Commun. Stat., Theor. Methods, A9(2), 213-230 (1980)

Pringle, R.; Rayner, A.: Generalized Inverses With Applications in Statistics., Hafner (1971)

Rao, C. R.: A Note On A Generalized Inverse Of A Matrix With Applications To Problems In Mathematical Statistics., J. R. Stat. Soc., Ser. B 24, 152-158 (1962)

Rao. C. R.: Projectors, Generalized Inverses and the BLUE's., J. R. Stat. Soc., Ser. B 36, 442-448 (1974)

Rao, C. R.; Mitra S. K.: Generalized Inverse of Matrices and Its Applications., New York, Wiley (1971)

Rao, C. R.; Mitra, S. K.: Theory and Application of Constrained Inverse of Matrices., Siam J. Appl. Math. 24, 473-488 (1973)

Rao, C. R.; Yanai, H.: Generalized Inverse of Linear Transformations: A Geometric Approach., Linear Algebra Appl. 66, 87-98 (1985)

SAS User's Guide Statistics, Cary, SAS Institute Inc. (1985).

Schmoyer, R. L.: Everyday Application of the Cell Means Model., Am. Stat. 38, 49-52 (1984)

Searle, S. R.: Linear Models., New York, Wiley (1971)

Searle, S. R.: Relationships Between the Estimable Functions of SAS GLM Output for Unbalanced Data and the Hypotheses Tested by Traditional-Style F-Statistics., SAS - SUGI4, 196-207 (1979)

Searle, S. R.: Restrictions and Generalized Inverses in Linear Models., Am. Stat. 38, 53-54 (1984)

Searle, S. R.; Speed, F. M.; Henderson, H. V.: Some Computational Model Equivalences in Analysis of Variance of Unequal Subclass-Numbers Data., Am. Stat. 35, 16-33 (1981)

Shanbhag, D. H.: On The Independence Of Quadratic Forms., J. R. Stat. Soc., Ser. B 28, 582-583 (1966)

Shanbhag, D. H.: Some Remarks Concerning Khatri's Result Result On Quadratic Forms., Biometrika 55, 593-595 (1968)

Smith, D. W.; Murray, L. W.: A Simplified Treatment of the Estimation of Parameters and Tests of Hypotheses iin Constrained Models With Unbalanced Data., Am. Stat. 37, 156-158 (1983)

Speed, F. M.; Hocking, R. R.: The Use of the R( )-Notation with Unbalanced Data., Am. Stat. 30, 30-33 (1976)

Speed, F. M.; Hocking, R. R.: A Characterization of the GLM Sums of Squares., SAS - SUGI 5, 215-218 (1980)

Speed, F. M.; Hocking, R. R.; Hackney, O. P.: Methods of Analysis of Linear Models with Unbalanced Data., J. Am. Stat. Assoc. 73, 105-112 (1978)

Tan, W. Y.: On the Distribution of Quadratic Forms In Normal Random Variables., Canadian J. Stat. 2, 241-250 (1977)

Tandon, P. K.; Lustick, R. R.: SAS Methods for Obtaining the Exact Hypotheses From Type I, II, III and IV Estimable Functions in Terms of Cell Means Models., SAS - SUGI 10, 1076-1083 (1985)

Yates, F.: The Analysis of Multiple Classifications With Unequal Numbers in the Different Classes., J. Am. Stat. Assoc. 29, 51-66 (1934)

Lecture Notes in Statistics

Vol. 1: R.A. Fisher: An Appreciation. Edited by S.E. Fienberg and D.V. Hinkley. XI, 208 pages, 1980.

Vol. 2: Mathematical Statistics and Probability Theory. Proceedings 1978. Edited by W. Klonecki, A. Kozek, and J. Rosiński. XXIV, 373 pages, 1980.

Vol. 3: B. D. Spencer, Benefit-Cost Analysis of Data Used to Allocate Funds. VIII, 296 pages, 1980.

Vol. 4: E. A. van Doorn, Stochastic Monotonicity and Queueing Applications of Birth-Death Processes. VI, 118 pages, 1981.

Vol. 5: T. Rolski, Stationary Random Processes Associated with Point Processes. VI, 139 pages, 1981.

Vol. 6: S. S. Gupta and D.-Y. Huang, Multiple Statistical Decision Theory: Recent Developments. VIII, 104 pages, 1981.

Vol. 7: M. Akahira and K. Takeuchi, Asymptotic Efficiency of Statistical Estimators. VIII, 242 pages, 1981.

Vol. 8: The First Pannonian Symposium on Mathematical Statistics. Edited by P. Révész, L. Schmetterer, and V. M. Zolotarev. VI, 308 pages, 1981.

Vol. 9: B. Jørgensen, Statistical Properties of the Generalized Inverse Gaussian Distribution. VI, 188 pages, 1981.

Vol. 10: A. A. McIntosh, Fitting Linear Models: An Application on Conjugate Gradient Algorithms. VI, 200 pages, 1982.

Vol. 11: D. F. Nicholls and B. G. Quinn, Random Coefficient Autoregressive Models: An Introduction. V, 154 pages, 1982.

Vol. 12: M. Jacobsen, Statistical Analysis of Counting Processes. VII, 226 pages, 1982.

Vol. 13: J. Pfanzagl (with the assistance of W. Wefelmeyer), Contributions to a General Asymptotic Statistical Theory. VII, 315 pages, 1982.

Vol. 14: GLIM 82: Proceedings of the International Conference on Generalised Linear Models. Edited by R. Gilchrist. V, 188 pages, 1982.

Vol. 15: K. R. W. Brewer and M. Hanif, Sampling with Unequal Probabilities. IX, 164 pages, 1983.

Vol. 16: Specifying Statistical Models: From Parametric to Non-Parametric, Using Bayesian or Non-Bayesian Approaches. Edited by J.P. Florens, M. Mouchart, J.P. Raoult, L. Simar, and A.F.M. Smith. XI, 204 pages, 1983.

Vol. 17: I. V. Basawa and D. J. Scott, Asymptotic Optimal Inference for Non-Ergodic Models. IX, 170 pages, 1983.

Vol. 18: W. Britton, Conjugate Duality and the Exponential Fourier Spectrum. V, 226 pages, 1983.

Vol. 19: L. Fernholz, von Mises Calculus For Statistical Functionals. VIII, 124 pages, 1983.

Vol. 20: Mathematical Learning Models – Theory and Algorithms: Proceedings of a Conference. Edited by U. Herkenrath, D. Kalin, W. Vogel. XIV, 226 pages, 1983.

Vol. 21: H. Tong, Threshold Models in Non-linear Time Series Analysis. X, 323 pages, 1983.

Vol. 22: S. Johansen, Functional Relations, Random Coefficients and Nonlinear Regression with Application to Kinetic Data. VIII, 126 pages. 1984.

Vol. 23: D. G. Saphire, Estimation of Victimization Prevalence Using Data from the National Crime Survey. V, 165 pages. 1984.

Vol. 24: T. S. Rao, M. M. Gabr, An Introduction to Bispectral Analysis and Bilinear Time Series Models. VIII, 280 pages, 1984.

Vol. 25: Time Series Analysis of Irregularly Observed Data. Proceedings, 1983. Edited by E. Parzen. VII, 363 pages, 1984.

ctd. on inside back cover

Lecture Notes in Statistics

Vol. 26: Robust and Nonlinear Time Series Analysis. Proceedings, 1983. Edited by J. Franke, W. Härdle and D. Martin. IX, 286 pages. 1984.

Vol. 27: A. Janssen, H. Milbrodt, H. Strasser, Infinitely Divisible Statistical Experiments. VI, 163 pages. 1985.

Vol. 28: S. Amari, Differential-Geometrical Methods in Statistics. V, 290 pages. 1985.

Vol. 29: Statistics in Ornithology. Edited by B.J.T. Morgan and P.M. North. XXV, 418 pages. 1985.

Vol. 30: J. Grandell, Stochastic Models of Air Pollutant Concentration. V, 110 pages. 1985.

Vol. 31: J. Pfanzagl, Asymptotic Expansions for General Statistical Models. VII, 505 pages. 1985.

Vol. 32: Generalized Linear Models. Proceedings, 1985. Edited by R. Gilchrist, B. Francis and J. Whittaker. VI, 178 pages. 1985.

Vol. 33: M. Csörgő, S. Csörgő, L. Horváth, An Asymptotic Theory for Empirical Reliability and Concentration Processes. V, 171 pages. 1986.

Vol. 34: D.E. Critchlow, Metric Methods for Analyzing Partially Ranked Data. X, 216 pages. 1985.

Vol. 35: Linear Statistical Inference. Proceedings, 1984. Edited by T. Caliński and W. Klonecki. VI, 318 pages. 1985.

Vol. 36: B. Matérn, Spatial Variation. Second Edition. 151 pages. 1986.

Vol. 37: Advances in Order Restricted Statistical Inference. Proceedings, 1985. Edited by R. Dykstra, T. Robertson and F.T. Wright. VIII, 295 pages. 1986.

Vol. 38: Survey Research Designs: Towards a Better Understanding of Their Costs and Benefits. Edited by R.W. Pearson and R.F. Boruch. V, 129 pages. 1986.

Vol. 39: J.D. Malley, Optimal Unbiased Estimation of Variance Components. IX, 146 pages. 1986.

Vol. 40: H.R. Lerche, Boundary Crossing of Brownian Motion. V, 142 pages. 1986.

Vol. 41: F. Baccelli, P. Brémaud, Palm Probabilities and Stationary Queues. VII, 106 pages. 1987.

Vol. 42: S. Kullback, J.C. Keegel, J.H. Kullback, Topics in Statistical Information Theory. IX, 158 pages. 1987.

Vol. 43: B.C. Arnold, Majorization and the Lorenz Order: A Brief Introduction. VI, 122 pages. 1987.

Vol. 44: D.L. McLeish, Christopher G. Small, The Theory and Applications of Statistical Inference Functions. 136 pages. 1987.

Vol. 45: J.K. Ghosh, Statistical Information and Likelihood. 384 pages. 1988.

Vol. 46: H.-G. Müller, Nonparametric Regression Analysis of Longitudinal Data. VI, 199 pages. 1988.

Vol. 47: A.J. Getson, F.C. Hsuan, {2}-Inverses and Their Statistical Application. VIII, 110 pages. 1988.